T0137256

Translational Systems Sciences

Volume 13

In 1956, Kenneth Boulding explained the concept of General Systems Theory as a skeleton of science. He describes that it hopes to develop something like a "spectrum" of theories—a system of systems which may perform the function of a "gestalt" in theoretical construction. Such "gestalts" in special fields have been of great value in directing research towards the gaps which they reveal.

There were, at that time, other important conceptual frameworks and theories, such as cybernetics. Additional theories and applications developed later, including synergetics, cognitive science, complex adaptive systems, and many others. Some focused on principles within specific domains of knowledge and others crossed areas of knowledge and practice, along the spectrum described by Boulding.

Also in 1956, the Society for General Systems Research (now the International Society for the Systems Sciences) was founded. One of the concerns of the founders, even then, was the state of the human condition, and what science could do about it.

The present Translational Systems Sciences book series aims at cultivating a new frontier of systems sciences for contributing to the need for practical applications that benefit people.

The concept of translational research originally comes from medical science for enhancing human health and well-being. Translational medical research is often labeled as "Bench to Bedside." It places emphasis on translating the findings in basic research (at bench) more quickly and efficiently into medical practice (at bedside). At the same time, needs and demands from practice drive the development of new and innovative ideas and concepts. In this tightly coupled process it is essential to remove barriers to multi-disciplinary collaboration.

The present series attempts to bridge and integrate basic research founded in systems concepts, logic, theories and models with systems practices and methodologies, into a process of systems research. Since both bench and bedside involve diverse stakeholder groups, including researchers, practitioners and users, translational systems science works to create common platforms for language to activate the "bench to bedside" cycle.

In order to create a resilient and sustainable society in the twenty-first century, we unquestionably need open social innovation through which we create new social values, and realize them in society by connecting diverse ideas and developing new solutions. We assume three types of social values, namely: (1) values relevant to social infrastructure such as safety, security, and amenity; (2) values created by innovation in business, economics, and management practices; and, (3) values necessary for community sustainability brought about by conflict resolution and consensus building.

The series will first approach these social values from a systems science perspective by drawing on a range of disciplines in trans-disciplinary and cross-cultural ways. They may include social systems theory, sociology, business administration, management information science, organization science, computational mathematical organization theory, economics, evolutionary economics, international political science, jurisprudence, policy science, socio-information studies, cognitive science, artificial intelligence, complex adaptive systems theory, philosophy of science, and other related disciplines. In addition, this series will promote translational systems science as a means of scientific research that facilitates the translation of findings from basic science to practical applications, and vice versa.

We believe that this book series should advance a new frontier in systems sciences by presenting theoretical and conceptual frameworks, as well as theories for design and application, for twenty-first-century socioeconomic systems in a translational and trans-disciplinary context.

More information about this series at http://www.springer.com/series/11213

David Rousseau • Jennifer Wilby
Julie Billingham • Stefan Blachfellner

General Systemology

Transdisciplinarity for Discovery, Insight
and Innovation

Foreword by Prof. Kyoichi Kijima

 Springer

David Rousseau
Centre for Systems Philosophy
Addlestone, Surrey, UK

Julie Billingham
Centre for Systems Philosophy
Addlestone, Surrey, UK

Jennifer Wilby
Centre for Systems Studies,
University of Hull
Kingston upon Hull, UK

Stefan Blachfellner
Bertalanffy Center for the Study
of Systems Science
Vienna, Austria

ISSN 2197-8832   ISSN 2197-8840   (electronic)
Translational Systems Sciences
ISBN 978-981-13-3817-5   ISBN 978-981-10-0892-4   (eBook)
https://doi.org/10.1007/978-981-10-0892-4

Printed on acid-free paper

This Springer imprint is published by the registered company Springer Nature Singapore Pte Ltd.
The registered company address is: 152 Beach Road, #21-01/04 Gateway East, Singapore 189721,
Singapore

Foreword

I am excited that this insightful and challenging research is now published as a volume of the Translational Systems Science Book Series. Over the past several years I have had a number of opportunities to learn about the progress of this ambitious research towards establising General Systemology, led by the team of authors. The research project is brought into shape here.

The pace of discovery in science is accelerating, and so also is the pace of innovation in the engineering disciplines, bringing new opportunities both to improve human quality of life and also to do so in sustainable ways. However, as science and technology advances, the specializations become more arcane to each other, raising the barriers to collaboration between the fields; slowing the flow of insights across disciplinary boundaries; and complicating the translation of urgent human and social needs into penetrating research projects, of scientific findings into effective technical solutions, and of solutions into beneficial practical use.

These challenges prompted the establishment of translational sciences, which aim to accelerate these transfers. Our increasing sensitivity to the systemic complexity of the world has resulted in the emergence of translational *systems* sciences, scientific disciplines that enhance the effectiveness of translational science by taking into account the systems perspective, or that show how to derive practical benefits from specialized systems theories.

Traditional translational science is grounded in establishing and maintaining a mapping between the advances, needs, resources, and values of different specialisms, connecting them while preserving their individuality. In this way the barriers between them are made permeable, and researchers and technologists can more readily collaborate with or be inspired by their colleagues from other disciplines. However, the challenge of keeping the barrier permeable is very substantial, because of the continuous introduction of new or modified concepts, terms, insights, and achievements on each side of the interdisciplinary interface.

This problem was already noted in the 1950s. Even then it was evident that many scientific activities were inefficient or duplicating efforts due to lack of effective communication across disciplines, and that progress in the human and social sciences was mismatched to our emerging technological powers, engendering substantial risks and problems for ecology, society, and individual well-being. This problematic trend is still continuing today, and the rising complexity of the sciences makes the need for more effective translational sciences more urgent than ever.

One approach to solving this problem, envisioned in the 1950s by Ludwig von Bertalanffy, Kenneth Boulding, Ralph Gerard, Anatol Rapoport, and others, was based on the idea of establishing a scientific general theory of systems, the principles and concepts of which would be a natural common ground shared by all the specialized disciplines. This would enable a new kind of translational capability. Rather than building specialized connections between the disciplines it would establish a common foundation the specialisms can all rest on, and which would not be weakened by specialized advances. Indeed, specialized insights would often be seen to be contextualized exemplars of general systems insights or even heralds of new general systems principles, thus strengthening and expanding the transdisciplinary foundation provided by the foundational scientific general theory of systems. Moreover, the knowledge of general systems concepts and principles would inspire and accelerate discovery and innovation within the specialized areas, in addition to making such innovation easier to communicate or engage with across the boundaries.

The discipline that is concerned with discovering, integrating, and putting into transdisciplinary use, scientific general systems insights and principles, is General Systemology, and I am delighted to provide a Foreword to the first book to employ that term in its title, and the first book to take the quest for a scientific general theory of systems beyond a grand vision and preliminary heuristics and put it on a principled scientific foundation. The work presented in the present volume has the potential to transform the effectiveness and efficiency of the translational sciences in general, and the translational systems sciences in particular.

Via this book the team of authors amply demonstrates both the necessity and the value of multidisciplinary collaboration, bringing together as they do expertise in science, mathematics, operational research, enterprise architecture, engineering, philosophy, management science, design science, and several systems sciences. Their collaboration has delivered breakthrough advances for the systems sciences, and opened up the way for a foundational scientific general systems theory to achieve the potential foreseen by the founders of the general systems movement.

I gladly and strongly recommend both their book and their collaborative style of working. The impact of this work on the translational systems sciences, and its value to our society, may take some time to become visible, but I am sure the positive contribution they have made toward bringing about a general systems transdiscipline cannot be doubted or underestimated. I keenly anticipate the developments that are sure to follow.

Tokyo, Japan   Kyoichi Jim Kijima, Dr.
2018   Professor Emeritus, Tokyo Institute of Technology
                                                       President of the International Society
                                                           for the Systems Sciences, 2007
                                                    President of Japan Society for Management
                                                                    Information, 2015–

Preface

Our knowledge of the world is fragmented, and given the complexity of the world this may be an enduring condition for humanity. However, the world itself is a unity, and thus we are always at risk of unintended negative consequences when we apply our limited knowledge. The growth of unsustainable practices, the increasing interconnectedness of the world, and the rising complexity of the systems we now seek to create, transform, or nurture have only served to increase this risk. Although a true unity of knowledge might be an unattainable goal, an increasing consilience of knowledge is not out of the question. One possible route to such consilience is offered by the vision of a general theory of systems. If everything in the world is a system or part of one, then general systems knowledge would not only be of transdisciplinary relevance, but afford deep insights about the interconnectedness of everything, and readily reveal to us important insights that cannot easily be seen from any specialized point of view.

The quest for a scientific general understanding of systems began with an inspiring vision presented in the 1950s by the founders of the general systems movement, led by Ludwig von Bertalanffy, Kenneth Boulding, Anatol Rapoport, and Ralph Gerard. Their motivation was not merely the scientific value of understanding something fundamental about the nature of the world, but rather the social value of recognizing that the world is more like a great organism than a great machine. Such a paradigm, they felt, would restore to humanity the dignity and value that mechanistic models tend to first ignore and then erode.

The promise of the founders' vision was clear but the path was not, and in the early years, conditions in academia were largely unfavorable for a project that would require such interdisciplinary collaboration. Consequently, dedicated research was limited and progress slow. In recent decades, however, conditions have become supportive of such projects, and the need for a general systems theory (GST) has become more urgent.

We, the present authors, came together in 2013 with a view to rekindle the enthusiasm for establishing a scientific general systems theory and operationalizing it as a transdisciplinary methodology for effective and efficient exploration, design, and theory building, and as a framework for visualizing the submerged land connecting our islands of knowledge. We felt that such a systems science framework would be essential for meeting the growing challenge of translating the progress in one area of knowledge to others who can benefit from it, and thus be a unifying framework not only for the sciences generally but for the Translational Systems Sciences particularly.

We felt that progress would require the collaboration of multiple specialists, and so we launched our *Manifesto for General Systems Transdisciplinarity* in 2015 not only to highlight the promise and key challenges of this research but to inspire and recruit researchers who would work together to accelerate progress toward attaining the vision of a practically useful general theory of systems. We were pleased to find our manifesto well received, and the present volume presents the work done by us and colleagues over the last few years. We are grateful to the organizations and individuals who have supported, inspired, and guided us as we developed and delivered a range of conference presentations, workshops, webinars, and publications. This communal effort is now starting to deliver practical results, and in this volume, we present not only the unifying conceptual framework that we called for in our *Manifesto* but also the first scientific systems principles that are of general relevance. We hope that this will prove to be the thin end of a thick and sturdy wedge, and that this book will serve as an invitation and a call for others to join the various teams working on establishing and leveraging a fully fledged general systems transdisciplinarity (GSTD).

In compiling the present volume we drew heavily on the contents of the publications and workshops we produced in the GSTD program, and we are greatly indebted to all who helped to shape the ideas presented here.

We express our gratitude to the many organizations who encouraged, supported, funded, promoted, and facilitated our work, in particular the International Society for the Systems Sciences, the Centre for Systems Studies in the University of Hull, the Centre for Systems Philosophy, the Bertalanffy Center for the Study of Systems Science, the International Council on Systems Engineering, the INCOSE Foundation, the Center for Agent-Based Social Systems Sciences in the Tokyo Institute of Technology, the European Meetings on Cybernetics and Systems, and the International Federation for Systems Research.

We thank the editors and reviewers of our papers published in the journals *Systems Research and Behavioural Science*, *Systema*, and *Systems*, and of our chapter published in the volume *Disciplinary Convergence in Systems Engineering Research*, for their helpful comments on drafts of these works.

We also thank colleagues who participated in or facilitated our workshops and webinars for providing us with stimulating debates, challenging questions, and general encouragement, who commented on drafts of our publications, who gave us advice and other help, and who engaged with us in building up a community of interest around the vision of General Systems Transdisciplinarity. In particular we thank (in no particular order) James Martin, Kyoichi (Jim) Kijima, Javier Calvo-Amodio, Debora Hammond, Richard Martin, Duane Hybertson, Gerald Midgley, Yasmin Merali, Mary Edson, Peter Caws, Alexander Laszlo, Ray Ison, Gary Metcalf, Ricardo Barrera, Raul Espejo, Andreas Hieronymi, Cecilia Haskins,

George Mobus, Janet Singer, Michael Singer, Peter Tuddenham, Pam Buckle, Gary Smith, Brigitte Daniel Allegro, Andrew Pickard, Siqi Wang, Sage Kittelman, Mike C. Jackson, Bill Schindel, John Kineman, Anand Kumar, Randall Russell, Robert Sherman, Swaminathan Natarajan, Chris Paredis, Paul Collopy, Mike Celentano, Mike Watson, Jean-Claude Roussel, Giles Hindle, Angela Espinosa, Robert Edson, Mario Bunge, and Len Troncale.

We owe you more than we can express, and we look forward to ongoing collaboration with you and new colleagues as together we develop General Systemology for the benefit of society and our world.

Surrey, UK   David Rousseau
Kingston upon Hull, UK   Jennifer Wilby
Surrey, UK   Julie Billingham
Vienna, Austria   Stefan Blachfellner

Contents

Abbreviations

ASEM	American Society for Engineering Management
AKG model	"Activity Scope – Knowledge Base – Guidance Framework" model of a discipline
BCSSS	Bertalanffy Center for the Study of Systems Science
CCP	Conservation of Properties Principle
EMCSR	European Meetings on Systems and Cybernetics
GSSPs	General Scientific Systems Principles
GSMs	General Systems Methodologies
GST	General Systems Theory
GST*	General Systems Theory, meaning the foundational theory of the nature of systems
GSTD	General Systems Transdisciplinarity
GSW	General Systems Worldview
IFSR	International Federation for Systems Research
INCOSE	International Council on Systems Engineering
ISSS	International Society for the Systems Sciences
PLT model	the "Principles-Laws-Theories" model of modern science
SRBS	Systems Research and Behavioral Science (journal)
SSPs	Scientific Systems Principles

Chapter 1
Introduction

Abstract The quest for a scientific general systems theory formally started in the 1950s, but progress has been slow. In this chapter we introduce General Systemology, and discuss its origins in the 1950s and its subsequent history in the general systems movement. We discuss its potential and challenges, and outline the key steps needed to establish it as a useful academic discipline.

Keywords Systemology · General systemology · General systems theory · GST · GST

1.1 Introducing "General Systemology"

This book is about recent developments in the philosophical and scientific quest to understand the nature of systems. Ever since Aristotle famously pointed out that a system is "a whole that is more than the sum of its parts", it has been recognized that 'system' is a special kind of thing or category of thought beyond other kinds of things and models. However it has only recently come to be appreciated how much more there is to the notion of 'system' than is given in Aristotle's dictum, and how important knowledge of that 'more' could be for science and society. The last century has seen not only the emergence of a strong scientific interest in systems, but also the rise of a "systems perspective", from which people see systems as significantly present in the world (and/or in thinking about the world), and hence they seek insight into the nature and workings of systems as a basic requirement for adequate description, analysis and practical engagement with the real world.

Although initial progress was slow the systems perspective has recently become a significant academic interest (Capra & Luisi, 2014; Hooker, 2011; Midgley, 2003). Increasingly, knowledge of systems is seen as presenting a paradigm for addressing complex problems, that is, those involving phenomena that cannot be adequately modelled using the classically powerful approaches based on reductionism and linear causal mechanisms (Dekkers, 2014; Mobus & Kalton, 2014). Additionally, it is ever more valued for its potential to support transdisciplinarity, i.e. the principles and models that characterise aspects of systemicity can be applied in multiple disciplines (for example the principles associated with stabilizing feedbacks (Wiener, 1961) and near-decomposable hierarchies (Simon, 1962)). The systems perspective is progressively seen as both necessary for understanding the complexity of the world in general, and as useful to researchers in a multitude of specialised fields.

A core claim under the systems perspective is that everything we encounter is a system or part of one. If this is true then 'being a system', i.e. having the attribute we might call 'systemness' or 'systemhood', or being something that is 'systemic', is a matter of considerable significance. But what is that significance? The full meaning of the term 'system' is not settled yet, but the term 'system' appears to be used somewhat like how we use the term 'energy', a general term for the something we can only know through specific instances. And just as coming to understand the nature of energy transformed our understanding of how specific things work and what particular kinds of change are possible, so too, perhaps, will understanding the nature of systems transform our understanding of the world as a grand scheme, and transform our understanding of our place and our potential within that scheme.

Over the last century many scientists and philosophers grappled with the problem of explaining the nature of systems, but the work that first became influential in the West was that of the Austrian biologist Ludwig von Bertalanffy (1901–1972). Von Bertalanffy coined the term "General Systems Theory" (GST) as an attempt to translate his original German term "Allgemeine Systemlehre". This term has no exact equivalent in English, but it means something like "an organized body of knowledge about systems that is of general (i.e. of transdisciplinary) significance". With hindsight it has become clear, as we discuss later on, that the term "GST" was neither an apt choice nor narrowly defined, and this has led to it acquiring dozens of different meanings in the literature, and entrenching confusions that a better terminology would eliminate. For this reason systems researchers have in recent years proposed the term "Systemology" (Pouvreau & Drack, 2007) to refer to the organized body of knowledge about systems, and "General Systemology" (ibid) to refer to the subset of systemology that represents the organised body of knowledge about the inherent nature of all systems, that is to say about what is essential to or universally true about systems. General systemology is thus especially concerned with those attributes that confer "systemhood" or "systemness" or "systemicity" on things that we recognize as systems, and how the combination of these universal attributes gives rise to the behaviours we see in specialized kinds of systems. For brevity this is sometimes referred to as knowledge about, or attributes of, "systems as such" or "systems as systems", distinguishing such knowledge from knowledge about specialized kinds of systems, which only applies to some systems or in some contexts. The terms 'systemology' and 'general systemology' thus capture two of the many meanings historically embraced by the term "GST". We will continue to disambiguate the concepts swirling around current uses of systems terminology as the book progresses, to provide a clearer foundation for advancing and discussing our knowledge of systems and its application.

General systemology is still in the early stages of its development, but like any other scientific discipline its scope would develop to include concepts, principles, theories, methods and practices, and hence be more than just a theory (or group of theories). The central theory of general systemology would be, as mentioned earlier,

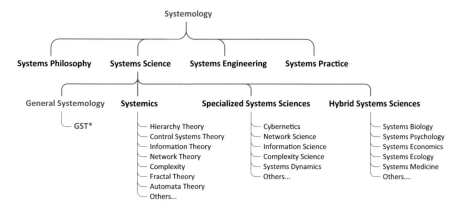

Fig. 1.1 General systemology in context

the one that explains the nature of systems.[1] That theory could aptly be called "GST", but to avoid confusion with the many other historical uses of the term "GST" we proposed referring to this theory (or group of theories) as GST* (pronounced "g-s-t-star") (Rousseau, Wilby, Billingham, & Blachfellner, 2015a, 2016a).

Systemology, in the sense just defined, is a broad field, and encompasses systems philosophy, systems science, systems engineering and systems practice. As will be explained later on, 'systems science' encompasses the discipline of general systemology (which includes the general theory about the nature of systems (GST*)), various specialised systems sciences (for example cybernetics, network science, information science, complexity science), and the hybrid systems sciences (which includes the disciplines dealing with the systemic aspects of specialised subject interests, for example systems biology, systems psychology etc.). The specialised systems sciences are grounded in a range of specialised systems theories collectively known as the "Systemics" (representing the collection of specialised theories dealing with particular aspects of systemic behaviour, for example hierarchy theory, control systems theory, automata theory, etc.). This too will be discussed in depth in further chapters.

The basic distinctions just enumerated are illustrated in Fig. 1.1.

With these basic distinctions in hand we can now embark on the exploration of the history, significance and potential of a general understanding of the nature of systems. We will in the course of this book present recent advances in the quest to establish general systemology, including foundational insights and breakthroughs that might open up new avenues for realising the potential of GST* to support transdisciplinarity for the benefit of humanity, as first envisioned by the founders of the general systems movement.

---

[1] There might also be others, for example theories about how to interpret the general theory under various contexts or worldviews, about how systems arose in nature in the first place (if indeed systems are emergent and not in some sense primordial), about the completeness of GST*, and about strategies for discovering general systems principles.

1.2   The Origin, History and Potential Significance of General Systemology

Systems have been subjects of philosophical interest for about 2500 years, beginning with the systemic models of the Pre-Socratic philosophers in ancient Greece, and famously expressed in Aristotle's famous dictum that in a system the whole is more than the sum of its parts (Metaphysics, Book 8.6.1045a:8–10, see (Aristotle, 2005, p. 126)). Systems have been studied scientifically for about 250 years, beginning with Étienne Bonnot de Condillac's Treatise on Systems (Condillac, 1749; Hine, 1979), in which he argued for philosophical systems to be subject to scientific standards of rationality and empirical verification. Today the study of systems is a rich field encompassing dozens of specialised systems theories and hundreds of methodologies (Midgley, 2003).

However the field is not yet unified because we are still lacking a general theory of systems. The existence, in principle, of a GST* was first suggested about a hundred years ago (Bogdanov, 1913; von Bertalanffy, 1932), but the quest for establishing it only took hold in the West after the middle of the last century, and this was largely due to the work and advocacy of Ludwig von Bertalanffy, who is now widely regarded as the founder of the "general systems movement". Starting in the 1930s, von Bertalanffy pointed out the isomorphic recurrence of systemic structures and processes in several specialized disciplines (for example hierarchies that support robustness and feedbacks that bring stability), and inferred that these isomorphisms denote the existence of general systems principles that could ground a GST*:

> Thus, [in principle] there exist models, principles and laws that apply to generalized systems, or their subclasses, irrespective of their particular kind, or the nature of their component elements, and the relations or "forces" between them. It seems legitimate to ask for a theory, not of systems of a more or less special kind, but of universal principles applying to systems in general. In this way we come to postulate a new discipline, called General System Theory. Its subject matter is the formulation and derivation of those principles which are valid for "systems" in general. A first consequence of the existence of general systems properties is the appearance of structural similarities or isomorphies in different fields (von Bertalanffy, 1956, p. 37, 1969, p. 32).

Inspired by von Bertalanffy's vision, Kenneth Boulding, Anatol Rapoport and Ralph Gerard, together with von Bertalanffy, founded the Society for the Advancement of General Systems Theory in 1954 with a core ambition to discover and leverage a GST* (Hammond, 2003, pp. 245–248). The Society was incorporated as the Society for General Systems Research (SGSR) in 1956, and the founders called for the development of a new discipline with "interdisciplinary" and "meta-scientific" aspects (von Bertalanffy, 1972, p. 421), grounded in a "General Systems Theory" that would encompass the principles underlying the systemic behaviours of all kinds of systems (Boulding, 1956; Gerard, 1964; Rapoport, 1986; von Bertalanffy, 1950a, 1969).

The founders believed that a GST* would support interdisciplinary communication and cooperation, facilitate scientific discoveries in disciplines that lack exact theories, promote the unity of knowledge, and help to bridge the divide between the naturalistic and the human sciences (von Bertalanffy, 1972, p. 413, 423–424; Rapoport, 1976; Laszlo, 1974, pp. 15–16, 19). The pioneers of general systems research saw this as a strategy and action plan for averting immanent social and environmental crises, and for opening up a pathway towards a sustainable and humane future (Hofkirchner, 2005, p. 1; Laszlo, 1972; Pouvreau, 2014, p. 180).

The SGSR lives on today as the International Society for the Systems Sciences (ISSS) (so renamed in 1988), but despite significant advances in the specialised systems sciences ("Systemics") the ambition to develop a GST* and leverage it for human and ecological benefit remains largely unfulfilled (Francois, 2007; Pouvreau, 2013, p. 864; Troncale, 2009, p. 553). Although general systemology is still a nascent discipline, much work towards it has been done since the founding of the ISSS, most notably by Len Troncale (ISSS President 1990–1991) and his colleagues (Friendshuh & Troncale, 2012; McNamara & Troncale, 2012; Troncale, 1978, 1984, 1985, 1986, 1988, 2001, 2006, 2017), who not only developed a substantial database of the isomorphic systems patterns that inspired the vision for a GST* but also worked out many interdependencies between them (which they call 'linkage propositions'). GST*'s development remains an active enterprise,, and recent times have seen an upsurge of interest in both GST* and General Systemology (Adams, Hester, Bradley, Meyers, & Keating, 2014; Arnold & Wade, 2015; Billingham, 2014a, 2014b; Denizan & Rousseau, 2014; Drack & Pouvreau, 2015; Drack & Schwarz, 2010; Green, 2015; Hofkirchner & Schafranek, 2011; Rousseau, 2015b; Rousseau et al., 2015a; Whitney, Bradley, Baugh, & Jr., 2015; Wolkenhauer & Green, 2013; Rousseau, Wilby, Billingham, & Blachfellner, 2015b, 2016b; Rousseau & Wilby, 2014, pp. 673–674); Billingham, 2014a; Rousseau & Wilby, 2014).

New attempts are also being made to bring GST* to bear on subjects as diverse as soil ecology (Lin, 2014), the mind-body problem (Rousseau, 2011), engineering of systems (Rousseau & Wilby, 2014), systems engineering (Adams et al., 2014), biology (Green, 2015), ethics (Rousseau, 2014d) and spirituality studies (Rousseau, 2014a). These efforts are distinct from new work specifically aimed at advancing GST*, which is the main focus of the present book.

For various reasons we discuss later on, the study of systems in general has not progressed as far or as fast as Von Bertalanffy envisaged, although the study of specific disciplinary systems has been highly productive. There has even been debate about whether a general theory exists, and whether Systemology is in fact any more than simply the sum of the systemics and the specialized systems sciences that grew out of them.

In this book we argue that the answer to both questions is yes, and that there are enormous benefits still to be gained from Von Bertalanffy's original vision. We explore each question in more detail. Then we consider the nature of disciplines and fields, and what would be needed for General Systemology to become established as one. Having shown that the existence of a General Theory is one of the stepping stones towards that goal, we then set about evaluating whether such a theory might exist, examining what it would look like, investigating how we might find it, extrapolating what its potential might be, and laying down foundations for how we can establish progress towards it. In the final chapter all this work will come together as we introduce newly discovered general scientific systems principles and discuss their practical value.

1.3   Key Steps Towards Systemology As a Field in Itself

Despite increasing academic interest in the systems perspective, Systemology is still an emerging field, and faces many obstructions to its unfolding. The field of systems is subject to many significant technical and political challenges, which will need to be overcome if it is to develop into a powerful, unifying field. We outline the steps needed below, and in Chaps. 2 and 3 offer ideas and concepts to support achieving this aim.

1.3.1   A Common Perspective on the Meaning of 'System'

The systems domain is ostensibly unified by a common subject matter (the "system" concept and its relevance to solving complex real world problems), but in practice this is only an administrative or political unity, and not a scientific one, because there is a considerable variety of perspectives within the systems community about the meaning of "system". For example, to some, "systems" are parts working together to perform a function of interest to humans (a typical systems engineering perspective), to others, "systems" are parts working together to establish an identity-preserving boundary (the organismic perspective) and to others a "system" is any collection of parts that we find a reason to consider in conjunction, treating it is a "system of interest" in which the interactions between the selected parts relate to the behaviours we are interested to study (the social constructivist perspective). To some, systems are present in the natural world as a subject of study (the critical realist perspective), while to others, systems concepts and models are merely useful constructs for thinking about human-centric experiences of complex situations and have no inherent truth-value (the post-modern perspective). In this way the system concept, which is the common denominator of the systems domain, is to a large degree also the basis of its fragmentation. Charles Francoise's two-volume *International Enclopedia of Systems and Cybernetics* (Francois, 2004) devotes 18 pages to listing variations on the system concept (2.6% of the total pages in the encyclopedia). This diversity of ideas around the system concept has been widely valued in the systems community as an antidote to intellectual hegemony in a nascent disciplinary field. However, the lack of such a common understanding is also problematic for the development of a disciplinary field. Without a stable concept of what "system" stands for, it is hard to make sense of and defend the claim that there is, or should be, a "systems discipline". An important step towards overcoming the systems domain's tentative status in the academy and becoming a unified disciplinary field in its own right, is therefore to develop a perspective on "system" that can reconcile the currently diverse perspectives, for example by accommodating them as special cases under a general definition.

1.3.2 Alignment of Terminology

The terminological challenges of the system domain extend well beyond the meaning of "system".

There is no standard nomenclature for referring to the components of the systems body of knowledge, or even the domain as a whole. Terms such as "systems thinking", "systems science", and "systems research" are in play as names for the "domain" but none of them are used consistently or universally.

A further problem is that the well-known term "GST" is polysemic, having many significant designations in the systems literature. In the work of Ludwig von Bertalanffy, the founder of GST, we already find the following 16 meanings: a foundational theory, a scientific discipline, a new philosophy of nature, a worldview, a paradigm, a methodological maxim, a new field, a new realm of science, an organized body of knowledge, a systems epistemology, a theory of organizational principles, a systems philosophy, a collection of isomorphies, a field of science, a value theory, a meta-discipline (for example von Bertalanffy, 1968, p. 33, 1972, p. 414, 1975, p. 12, 1976, pp. xix–xxiii). The polysemy significantly limits scholarly exchange and collaboration in efforts towards the establishment of a GST. For example, if someone should now ask for criteria by which we might recognize a GST (as discussed by Len Troncale in Troncale, 1984), how shall we know which aspect of GST to suggest criteria for? If someone argues for the importance of "GST", or says that they are working on or using "GST", how can we know what they are meaning to refer to? If we are to undertake disciplined collaborative research towards a unifying theory for the systems field, this problem must be resolved as a first priority. Without it researchers cannot productively collaborate, and a unifying theory will continue to elude us. As mentioned above we have proposed the term GST* to refer to the theory about the fundamental nature of systems as systems, but beyond this other terms have to be established to become conventional ways of referring to the other meanings historically attached to the term "GST".

1.3.3 A Unified Vision for the Systems Field

The systems community is currently represented by a diversity of perspectives and methodologies (Dekkers, 2014; Midgley, 2003; Skyttner, 2006), a variety of views about its significance and potential (Denizan & Rousseau, 2014; Dubrovsky, 2004; Rousseau & Wilby, 2014; Warfield, 2003), and multiple perspectives on the possibility and viability of a unifying framework (Boulding, 1956; Gaines, 1979; Midgley, 2001; Rousseau, 2014e).

Such variety is not a serious problem per se, as both new and established disciplines go through cycles of critical reflection, fragmentation, and unification as they evolve. Current examples include, amongst emerging disciplines, the Science of Team Science (Falk-Krzesinski et al., 2011) and Transdisciplinary Action Research (Stokols, 2006), and, amongst recently established disciplines, Systems Engineering (Collopy, 2012; Pennock & Wade, 2015; Soban, Price, & Hollingsworth, 2012) and Evolutionary Developmental Biology (Green & Wolkenhauer, 2013; Jaeger, Laubichler, & Callebaut, 2015).

However, in the case of the systems field this variety is rather surprising. Systems science is not a new field – the first scientific work on systems appeared in the eighteenth century (Condillac, 1749) – but nevertheless systems science has made scant progress towards a unifying framework (Francois, 2006), despite this already being called for early on in the twentieth century (Bogdanov, 1913; von Bertalanffy, 1932) and many re-iterations of it since (Drack, 2009; Drack & Pouvreau, 2015; Drack & Schwarz, 2010; Friendshuh & Troncale, 2012; Hofkirchner & Schafranek, 2011; Klir, 1969; Pickel, 2007; Pouvreau, 2011; Sirgy, 1988; Troncale, 1984, 2009, von Bertalanffy, 1955, 1969). One reason for this slow progress may be that the community of researchers working towards developing unifying frameworks has always been rather small, and funding for such work has never been substantial (Drack & Schwarz, 2010; Rousseau & Wilby, 2014). This lack of attention and funding has slowed progress towards the systems field finding its unifying framework and, by implication, the prospect of the systems field making a significant contribution to solving the serious systemic challenges facing present-day socio-ecological systems.

1.3.4   A Disciplinary Map of the Systems Field

There is no overall view of the structure of the domain of systems as a disciplinary field, so the existing body of knowledge cannot be evaluated in terms of its completeness, internal relationships and potential; − it is a body of knowledge but not an academically organized one.

Several overviews of the systems domain have been presented by systems thinkers, but these have typically focused on viewing the domain from the perspective of its evolution, that is, by tracing the temporal origins of specific systems concepts, theories, perspectives or practices, and trying to show in each case how the emergence of each component was influenced by existing components, specific individuals, and strands of thought from beyond the domain of systems.. Examples of such "maps" are given in Troncale (1978), Van Gigch & Kramer (1981), Troncale (1988), Laszlo & Laszlo (2003), Graf (2006), Ison (2008), Ramage & Shipp (2009), Wright (2012) and Castellani (2012). For easy reference these examples are collated in

Appendix 2 of the open-access paper by (Rousseau, Wilby, et al., 2016a), located here: www.systema-journal.org/article/view/402/362.

Maps such as these give a sense of the evolutionary development and rich scope of the domain of systems, but these schemas also emphasize how deeply fragmented the systems domain is. They give no sense of the domain of systems as (potentially) a disciplinary field, a field (as will be discussed in later on) being a set of disciplines that represent specialized perspectives on a common subject matter, and that serve a broadly common purpose in relation to that common subject matter.

The closest we have come, historically, to a principled view of the structure of the systems domain regarded as a field was with John van Gigch and Nic Kramer's paper "Taxonomy of Systems Science" (van Gigch & Kramer, 1981). They suggested a classification in terms of two dichotomies (ontological/conceptual and theory/application), giving four possibilities. On this basis they proposed a division of systems science into four "branches" respectively representing:

- theories that treat reality as a system of systems (the "ontological-theoretical" branch, which they characterize as Systems Philosophy);
- theories that treat systems as formal models (the "conceptual-theoretical" branch, which they characterize as Axiomatic Systems Science);
- applications that treat a system as an organism (the "applied-ontological" branch, which they characterize as Living Systems Theory); and
- applications that treat problems as systems (the "applied-conceptual" branch, which they characterize as Systems Methodology).

Regrettably, in their paper they did not go on to use this framework to classify actual disciplinary components but only listed the individuals who were representative contributors to each of the "branches". Moreover, neither they nor others have attempted to develop this model any further. Thus, as it stands, this model, like the other "maps" referred to earlier, also does not provide a basis for assessing the completeness, prospects, unity or potential future value of the systems "field".

These historical systems domain maps depict the systems domain as fragmented into a collection of sub-specializations, united only by the common denominator "systems". By presenting the systems domain as an evolutionary trajectory of a plurality of approaches to investigation and intervention as is done at present, these maps emphasize and may entrench the fragmentation of the domain. If the systems community is to address this problem effectively, then it must begin to develop a perspective of the domain understood as a unified discipline or field, and come to a common understanding of the basis on which it can be unified.

1.3.5   A Stable Presence in Academic Institutions

The potential presence of the systems field as a distinct enterprise within academia is often questioned, and this has impeded the field's progress. The existing specialized systems theories and methodologies are valued by the disciplines that employ

them, but in the absence of a unifying framework and indigenous theoretical foundation for the systems sciences, the presence in academia of systems science as a
discipline in its own right is constantly under threat. Centres and Units promoting
systems science as such typically have a short lifetime in Universities, soon falling
foul of departmental re-organizations in which the specialized systemists are moved
to the orthodox departments that typically employ them (Warfield, 1990). In this
way we keep returning to a situation where we have for example systems biologists
in biology departments and systems management scientists in business schools but
typically no unit responsible for promoting and developing systems science as such.
This is a vicious cycle that undermines the ability of systems science to attract funding, recruit students, build credibility, and expand its ability to contribute positively
to addressing urgent social and environmental challenges. This has a knock-on
effect of weakening the academic standing of other disciplines in systemology such
as systems philosophy, systems engineering and systems practice. The key driver of
this cyclical fragmentation is the absence of a GST* that can unify the systems field
in a profound way. Without it, the field is largely just a collection of specializations,
creating the impression that little is gained by bringing them together under a common administration, and little is lost by allocating them to specialised departments
that employ them in isolation. However this fragmentation breaks down the lines of
communication between the specialist systemists, further impeding the discovery
and evolution of general systems insights.

In the next sections, we will address steps that can be taken to help establish a
GST* that can not only unify the specialized systems theories but also bring together
under a common enterprise-vision the disparate pockets of interest and activity that
exist in academia and industry today.

1.4   Key Steps Towards Establishing a General Theory
of Systems

1.4.1   Moving Beyond Heuristic Methodologies

For many systems methodologies there is no or limited grounding of the methodology in scientific theory, and this makes it difficult to defend their scientific merit.
Other disciplines have been in this situation and overcome it over time, for example
the development of scientific foundations enabled the separation of chemistry from
alchemy, astronomy from astrology, and psychology from phrenology. Without a
similar revolution in Systemology it will not be able to become an established academic enterprise. Some researchers have begun to explore foundations for such a
revolution in specialized areas of systemology such as service systems science
(Wilby, Macaulay, & Theodoulidis, 2011) and called for increased research regarding transdisciplinarity (Wilby, 2011), but much still remains to be done.

Many of the methodologies used in systemology appear to have some practical value, but without a grounding in a scientific theory their principles are only heuristics. The absence of a theoretical basis means that when these methods fail there is no principled basis on which to assess what went wrong, and this casts doubts on whether their successes were predicated on the use of the methodology or merely serendipitous. This limits the potential of these methodologies to be improved.

1.4.2   Establishing a Framework for Discovering New Systemics

There is as yet no principled framework for discovering new specialized systems theories. The specialized systems theories we have so far were developed in orthodox fields such Chemistry and Telecommunications Engineering and only afterwards recognized as potential Systemics and generalised (Minati, 2006). As things stand, we have no means for assessing the theoretical completeness of the specialized systems theories and for focusing developmental research programs on important challenges. A foundational general theory of systems could provide such a framework, just as the early Periodic Table of Elements suggested the existence of elements yet to be discovered.

1.4.3   Developing a Scientific Theory About the Nature of Systems

The systems movement was founded on the premise that a general theory of systems could be developed (Boulding, 1956; Rapoport, 1968; von Bertalanffy, 1950a, 1969), but, as reported earlier, limited progress has been made so far in developing it (Francois, 2007).

We presently have a vast array of heuristic principles expressing general sentiments about systems (e.g.. Augustine, 1987; Gall, 1978; Mobus & Kalton, 2014, pp. 17–29; Senge, 1990; Sillitto, 2014, pp. 33–38; Skyttner, 2006, pp. 99–105), but little connection between them and not much by way of principles that can be used to build scientific theories about the general nature of systems.

Making progress towards a more complete general theory of systems is crucial for the academic unity, credibility and advancement of the systems field. As discussed above, this means moving towards having scientific models that can reconcile the different perspectives on the nature of 'system' in a compelling manner. To support such a scientific unity the subject matter must be defined in terms of a theoretical framework that has explanatory and/or predictive value. Such a scientific general theory provides a conceptual and explanatory foundation on the basis of which the discipline or field can grow as a scientific endeavor of increasing epistemic and empirical competence.

In the life of a discipline or field the transition from viewing its subject matter merely in terms of descriptive models and theories to being able to represent it in terms of explanatory/predictive theories is of crucial significance. It is well known from the history of science that general theories such as Newton's Laws of Mechanics, Mendeleev's Periodic Table of the Chemical Elements, Lyell's Principles of Geology, and Darwin's Theory of Biological Evolution, transformed their respective disciplinary fields by (a) unifying hitherto fragmented areas of study under a common conceptual and explanatory framework, and (b) rapidly opening up new avenues to scientific discovery.

In the case of the systems domain, the sought-for scientifically-unifying theory would be the "General Systems Theory" (GST*) as originally envisioned by Ludwig von Bertalanffy (von Bertalanffy, 1956, p. 37, 1969, p. 32). Von Bertalanffy proposed that structures and behaviors that recur isomorphically across kinds of systems indicated the existence of general systems principles that would underpin the formulation of general systems laws that could be applied in diverse disciplines for problem solving, modelling, and design (von Bertalanffy, 1976, pp. xix–xxiii). Although Len Troncale long ago identified more than 50 candidate isomorphies (Troncale, 1978), and his list has since grown to over a hundred (Friendshuh & Troncale, 2012), we so far have almost no systems principles that could count as general and practically useful. However, that said, we do have a few, including von Bertalanffy's proposal that there are no closed systems in nature (von Bertalanffy, 1950b) and Herbert Simon's proposal that all complex systems are near-decomposable hierarchies (Simon, 1962). We have nothing like a "periodic table of systems"; the closest we have come is Kenneth Boulding's hierarchy that orders systems in terms of the complexity of their behavior and the models and disciplines used to describe them (Boulding, 1956), but this model has many shortcomings (Billingham, 2014a; Mingers, 1997; Wilby, 2006).

The key advances toward a GST* seem mostly to have been made long ago, and general systems research has been a minority endeavor for the last 30 years. In reality, it was the practical offshoots of theories about individual isomorphies that took precedence, resulting in advances in Information Theory, Cybernetics, Organization Theory, Control Theory, Management Science, and so on. This pragmatic focus produced progress at a high cost, for "it left these theories together with the possibility of a "GST" philosophically immature" (Flood & Robinson, 1989, p. 63).

1.5   Recent Developments in General Systemology

In the light of the compelling vision, unfulfilled promise and rising need for GST* the present authors formed a team in late 2014 to address this issue, because of our belief that the task is important, achievable and urgent. Specifically, we presented supporting arguments (Billingham, 2014a; Rousseau, 2014b, 2014c, 2015a;

Rousseau & Wilby, 2014; Rousseau et al., 2015a; Rousseau, Wilby, Billingham, & Blachfellner, 2015c, 2015d; Wilby et al., 2015), that:

- the systems field cannot become an established academic discipline without developing a unifying framework grounded in a general theory of systems;
- that a unifying framework for the systems field exists in principle and that its development is a practical prospect; and
- that such a unifying framework would support the development of powerful and useful systemic methodologies for discovery, insight, innovation, intervention, management, control and engineering in all branches of science.

In the light of this position we launched in 2015 a Manifesto for General Systems Transdisciplinarity (Rousseau et al., 2015b; Rousseau, Wilby, et al., 2016b), calling for renewed efforts toward the development of a foundational general systems theory for the systems field, and the development of methodologies and perspectives that would put it to practical use and fulfil the potential of the systems perspective. Our vision was for not only advancing the foundations and content of GST* but to establish its application as a vibrant new disciplinary activity, which we dubbed "General Systems Transdisciplinarity (GSTD)". The present book presents to a large degree the outcome of the work undertaken and advocated by the GSTD team.

We advocated the GSTD programme to consider anew questions such as:

- What is "GST"?
- How might it fit into the "systems field"?
- What would it look like?
- Does it exist in principle? Under what perspective(s)?
- How might we discover/develop it?
- What might its potential be? Would it have any distinctive powers?
- How can we support progress towards establishing it?
- What can we discover if we take on board recent developments in science and the philosophy of science and apply this to what we know about systems?

We published an extensive structured list of questions to be pursued as a research agenda for GSTD in 2016 (Rousseau, Blachfellner, Wilby, & Billingham, 2016a).

In our Manifesto we proposed that progress towards establishing a valuable and competent General Systemology can be made by focussing on the development of:

- a **General Systems Worldview (GSW)** that is informed by our best scientific knowledge, by new discoveries in systems science, by advances in general systems research, and by the debate about the unity of science and the plurality of perspectives employed in systems thinking and practice.
- a **General Systems Theory (GST*)** that includes:

  - an ontology of systems that can be used to describe systems and classify them in an unambiguous way;
  - models that characterize the conditions and processes that support the evolution, persistence or degradation of systems; and

- principles and theories that explain the mechanisms that underpin the evolution persistence or degradation of systems.

- **General Systems Methodologies** (**GSMs**) that can leverage GST* under the guidance of the GSW to:

  - extend and refine GST*, the GSW and the methods of General Systemology;
  - discover new Theoretical Systemics, i.e. specialised theories about kinds of systemic structures, processes, behaviours, etc., or enhance existing ones;
  - discover new Methodological Systemics, i.e. specialised methods for systemic research, design, engineering, management, education etc., or enhance existing ones; and
  - support exploratory science in all areas of scientific inquiry.

- **General Systems Transdisciplinarity** (**GSTD**) that employs the GSMs to address the looming and present crises facing human civilization; and to contribute to the building of a thriving future world.

The present book presents the development and outcomes of the GSTD programme and sets it in the context of historical and current systems research.

1.6   Structure of the Book and Its Developmental Context

The content of the book addresses in a systematic way the foundational challenges we outlined in our initial Research Agenda, exemplified by the questions listed in the previous section. The chapters follow the sequence of our journey through this interesting terrain, and we think it is important to acknowledge here the contribution these activities, individuals and organizations have to the chapters presented here.

In Chap. 1 we give an overview of the history, status and challenges of the quest for a general theory of systems, and it reflects the background research we did for our Manifesto. Some of this material was discussed and developed through:

- workshops hosted by the International Federation for Systems Research (IFSR) (Billingham, 2014b; Wilby et al., 2015), the International Council on Systems Engineering (INCOSE) (Rousseau, 2015b; Rousseau, Wilby, Billingham, & Blachfellner, 2015a, 2015c) and the International Society for the Systems Sciences (ISSS) (Rousseau, Wilby, Billingham, & Blachfellner, 2015d)
- conference papers presented at the European Meetings on Systems and Cybernetics (EMCSR) (Billingham, 2014a; Denizan & Rousseau, 2014; Rousseau, 2014a; Wilby, 2014) and the ISSS (Rousseau, 2014b; Rousseau, Wilby, Billingham, & Blachfellner, 2015b); and
- papers published in the journals Systems Research and Behavioral Science (SRBS) (Rousseau, 2014b), and Systema (Rousseau, Blachfellner, Wilby, & Billingham, 2016; Rousseau, Wilby, Billingham, & Blachfellner, 2016b)

In Chap. 2 we expose and clear up key terminological issues (including the multiple meanings of the term "general systems theory"), and develop a generic map of the structure of a disciplinary field so that it would be possible to assess the maturity of the systems field, identify priorities for developmental research, and discuss the place and function of a GST within the wider field of Systemology. We introduce the term GST* to designate the foundational theory about the nature of systems, in order to disambiguate this theory from other uses of the term GST. This chapter revises and extends the work mentioned earlier (especially Wilby et al., 2015) and also draws significantly on a paper published in Systema (Rousseau, Wilby, Billingham, & Blachfellner, 2016a).

In Chap. 3 we delve into the subject of transdisciplinarity, disambiguating it from other kinds of disciplinarity. We develop a vision for the potential scope and range of GSTD, including its potential for problem-solving research and for supporting translational science. We show that the challenge of translational science has a much wider scope than is usually supposed, and discuss how GSTD can contribute to the currently emerging translational systems science. This chapter revises and extends some of the aforementioned early work and also draws on:

- conference presentations given at the 9th International Systems Sciences Symposium: Translational Systems Science (Rousseau, Billingham, Wilby, & Blachfellner, 2016a) and at the ISSS (Wilby, 2016); and
- papers published in SRBS (Rousseau & Wilby, 2014) and in Systema (Rousseau, Wilby, Billingham, & Blachfellner, 2016c).

In Chap. 4 we discuss the historical debate about the potential existence of a GST, and build a case for its potential existence and practical value. We develop a model of the "General Systems Worldview" (GSW), and relate it to the general model of a worldview we started to develop in Chap. 3. We show that the assumptions of the GSW entail the potential existence of a practically valuable GST*. We discuss how the GSW can guide us to the discovery of general systems principles, and argue that both the GSW and GST are needed for the development of GSTD. This chapter revises and extends some of the aforementioned early work, builds on the previous chapters, and draws additionally on:

- a conference presentation given at the ISSS (Rousseau, 2016a);
- a workshop hosted by INCOSE (Rousseau, 2018c), and
- papers published in SRBS (Rousseau, 2014d; Rousseau & Wilby, 2014), the Proceedings of the IFSR (Wilby et al., 2015), the Journal for the Study of Spirituality (Rousseau, 2014c), and Systema (Rousseau, Billingham, Wilby, & Blachfellner, 2016c).

In Chap. 5 we develop a detailed model of how the knowledge base of a discipline evolves, and of the role of general theories within that. We use this model to develop a detailed model of the components of a GST, its developmental stages, and how it would underpin advances in the other aspects of the knowledge base of systems science. This chapter revises and extends some of the aforementioned early work (especially Billingham, 2014a), builds on the previous chapters, and draws additionally on:

- workshops hosted by the IFSR (Edson et al., 2016) and INCOSE (Rousseau, 2017b); and
- a paper published in Systema (Rousseau, Billingham, Wilby, & Blachfellner, 2016b).

In Chap. 6 we develop a detailed discussion about the nature and evolution of general principles in science, and apply these insights and the work discussed in earlier chapters to develop three general scientific systems principles, and discuss some of the practical implications of these principles. This chapter revises and extends some of the aforementioned early work, draws together many threads from the work in earlier chapters, and draws additionally on:

- presentations given at meetings of the ISSS (Rousseau, 2016b), the Centre for Systems Studies in the University of Hull (Rousseau, 2016c), the National Science Foundation (Rousseau, 2017a), the School of Mechanical, Industrial and Manufacturing Engineering, Oregon State University (Rousseau, 2017h), the annual Conference of Systems Engineering Research (CSER) (Rousseau, 2017k), the INCOSE International Workshops 2017 and 2018 (Rousseau, 2017j, 2018a, 2018b) and the INCOSE International Symposium 2017 (Rousseau, 2017c);
- workshops hosted by INCOSE (Rousseau, 2017e, 2018e, 2018d), the EMEA Sector of INCOSE (Rousseau & Smith, 2017) and by the American Society for Engineering Management (ASEM) (Rousseau, Calvo-Amodio, & Barca, 2017);
- webinars presented to the INCOSE Crossroads of America regional chapter (Rousseau, 2017l), the NASA Systems Engineering Research Consortium (Rousseau, 2017d), and the systems engineering communities of practice in MTSI Inc. (Rousseau, 2017g) and in Rolls-Royce (Rousseau, 2017m); and
- papers published in the journals SRBS (Rousseau, 2017f) and Systems (Rousseau, 2017i), and a book chapter published in the edited volume "Disciplinary Convergence in Systems Engineering Research" (Rousseau, 2018f).

1.7   Summary

In this chapter we introduce General Systemology, and discuss its origins in the 1950s and its subsequent history in the general systems movement. We discuss its potential and challenges, and outline the key steps needed to establish it as a useful academic discipline. We argue that the key enablers of progress would be establishing a common perspective on the meaning of the term 'system', consolidating the current diversity of conflicting terminologies, establishing a unified vision of the field's scope and potential, developing a disciplinary map of the systems field, establishing a secure presence in academia, moving beyond heuristic methodologies to scientific ones, developing a systematic framework for discovering new specialised systems theories, and developing a scientific theory of the general nature of systems.

References

Adams, K. M., Hester, P. T., Bradley, J. M., Meyers, T. J., & Keating, C. B. (2014). Systems theory as the foundation for understanding systems. *Systems Engineering, 17*(1), 112–123.

Aristotle. (2005). *Metaphysics – Aristotle*. LLC, South Dakota: NuVision Publications.

Arnold, R. D., & Wade, J. P. (2015). A definition of systems thinking: A systems approach. *Procedia Computer Science, 44*, 669–678.

Augustine, N. R. (1987). *Augustine's laws*. New York: Penguin.

Billingham, J. (2014a). GST as a route to new systemics. Presented at the 22nd European Meeting on Cybernetics and Systems Research (EMCSR 2014), 2014, Vienna, Austria.' In J. M. Wilby, S. Blachfellner, & W. Hofkirchner (Eds.), *EMCSR 2014: Civilisation at the Crossroads – Response and Responsibility of the Systems Sciences, Book of Abstracts* (pp. 435–442). Vienna: EMCSR, 2014.

Billingham, J. (2014b). In Search of GST. Position paper for the 17th Conversation of the International Federation for Systems Research on the subject of 'Philosophical Foundations for the Modern Systems Movement', St. Magdalena, Linz, Austria, 27 April – 2 May 2014. (pp. 1–4).

Bogdanov, A. A. (1913). *Tektologiya: Vseobschaya Organizatsionnaya Nauka [Tektology: Universal Organizational Science] (3 volumes)*. Saint Petersburg, Russia: Semyonov' Publisher.

Boulding, K. E. (1956). General systems theory – The skeleton of science. *Management Science, 2*(3), 197–208.

Capra, P. F., & Luisi, P. L. (2014). *The systems view of life: A unifying vision*. Cambridge: Cambridge University Press.

Castellani, B. (2012). *Complexity map*. Retrieved 15 March 2015, from http://sacswebsite. blogspot.co.uk/2012/11/new-version-of-complexity-map.html

Collopy, P. D. (2012). A research agenda for the coming renaissance in systems engineering. In *American Institute of Aeronautics and Astronautics Symposium* (pp. 799–801). Reston, VA/ Nas hville, TN.

Condillac, E. B. de. (1749). Traité des Systèmes, Ou l'on en démêles les inconveniens et les avantages.

Dekkers, R. (2014). *Applied systems theory* (2015th ed.). New York: Springer.

Denizan, Y., & Rousseau, D. (2014). Bertalanffy and beyond: Improving systemics for a better future. A Symposium of the EMCSR 2014, 22–25 April. In *Civilisation at the crossroads: Response and responsibility of the systems sciences* (pp. 409–410). Vienna: BCSSS.

Drack, M. (2009). Ludwig von Bertalanffy's early system approach. *Systems Research and Behavioral Science, 26*(5), 563–572.

Drack, M., & Pouvreau, D. (2015). On the history of Ludwig von Bertalanffy's "General Systemology", and on its relationship to cybernetics – part III: convergences and divergences. *International Journal of General Systems, 44*(5), 523–5571.

Drack, M., & Schwarz, G. (2010). Recent developments in general system theory. *Systems Research and Behavioral Science, 27*(6), 601–610.

Dubrovsky, V. (2004). Toward system principles: General system theory and the alternative approach. *Systems Research and Behavioral Science, 21*(2), 109–122.

Edson, M. C., Buckle, P., Ferris, T., Hieronymi, A., Ison, R., Metcalf, G., et al. (2016). Systems research: A foundation for systems literacy. In G. Croust (Ed.), *Proceedings of the eigh-*

teenth conversation of the International federation for systems research, 3–8 April 2016, St. Magdalena, Linz, Austria. Linz, Austria: SEA-Publications, Johannes Kepler University.

Falk-Krzesinski, H. J., Contractor, N., Fiore, S. M., Hall, K. L., Kane, C., Keyton, J., et al. (2011). Mapping a research agenda for the science of team science. *Research Evaluation, 20*(2), 145–158.

Flood, R. L., & Robinson, S. A. (1989). Whatever happened to general systems theory? In R. L. Flood, M. C. Jackson, & P. Keys (Eds.), *Systems prospects* (pp. 61–66). New York: Plenum.

Francois, C. (Ed.). (2004). *International Encyclopedia of systems and cybernetics.* Munich, Germany: Saur Verlag.

Francois, C. (2006). Transdisciplinary unified theory. *Systems Research and Behavioral Science, 23*(5), 617–624.

Francois, C. (2007). *Who knows what general systems theory is?* Retrieved January 31, 2014, from http://isss.org/projects/who_knows_what_general_systems_theory_is

Friendshuh, L., & Troncale, L. R. (2012). Identifying fundamental systems processes for a general theory of systems. In *Proceedings of the 56th annual conference, International Society for the Systems Sciences (ISSS)*, July 15–20, San Jose State University, 23 pp.

Gaines, B. R. (1979). General systems research: quo vadis? *General Systems Yearbook, 24*, 1–9.

Gall, J. (1978). *Systemantics: How systems work and especially how they fail.* New York: Pocket Books.

Gerard, R. W. (1964). Entitation, Animorgs and other systems. In M. D. Mesarović (Ed.), *Views on general system theory: Proceedings of the 2nd systems Symposium at case institute.* New York: Wiley.

Graf, H. G. (2006). *Systems thinking and practice map.* St. Gallen Zentrum for Zukunftsforschung. Retrieved from http://www.sgzz.ch/en/?Systems_Thinking_Practice:Virtual_Map

Green, S. (2015). Revisiting generality in biology: Systems biology and the quest for design principles. *Biology & Philosophy, 30*(5), 629–652.

Green, S., & Wolkenhauer, O. (2013). Tracing organizing principles: Learning from the history of systems biology. *History and Philosophy of the Life Sciences, 35*, 553–576.

Hammond, D. (2003). *The science of synthesis: Exploring the social implications of general systems theory.* Boulder, CO: University Press of Colorado.

Hine, E. M. (1979). *A critical study of Condillac's: Traite des systemes.* The Hague, The Netherlands: Martinus Nijhoff.

Hofkirchner, W. (2005). Ludwig von Bertalanffy, Forerunner of evolutionary systems theory. In *The new role of systems sciences for a knowledge-based society, Proceedings of the First World Congress of the International Federation for Systems Research*, Kobe, Japan, CD-ROM (isbn 4-903092-02-X) (Vol. 6).

Hofkirchner, W., & Schafranek, M. (2011). General system theory. In C. A. Hooker (Ed.), Vol. 10: *Philosophy of complex systems* (1st ed., pp. 177–194). Amsterdam, The Netherlands: Elsevier BV.

Hooker, C. (Ed.). (2011). *Philosophy of complex systems.* North Holland (Elsevier): Oxford, UK.

Ison, R. L. (2008). Systems thinking and practice for action research. In P. W. Reason & H. Bradbury (Eds.), *The sage handbook of action research participative inquiry and practice* (2nd ed., pp. 139–158). London: Sage.

Jaeger, J., Laubichler, M., & Callebaut, W. (2015). The comet cometh: Evolving developmental systems. *Biological Theory, 10*(1), 36–49.

Klir, G. J. (1969). *An approach to general systems theory.* New York: Van Nostrand Reinhold.

Laszlo, E. (Ed.). (1972). *The relevance of general systems theory.* New York: George Braziller.

Laszlo, E. (1974). *A strategy for the future.* New York: Braziller.

Laszlo, E., & Laszlo, A. (2003). The systems sciences in service of humanity. In F. Parra-Luna (Ed.), *Systems science and cybernetics* (pp. 32–59). Oxford, UK: EOLSS Publishers.

Lin, H. (2014). A new worldview of soils. *Soil Science Society of America Journal, 78*(6), 1831.

McNamara, C., & Troncale, L. R. (2012). SPT II.: How to find & map linkage propositions for a general theory of systems from the natural sciences literature. In *Proceedings of the 56th annual conference, International Society for the Systems Sciences (ISSS)*, July 15–20, San Jose State University, 17 pp.

Midgley, G. (2001). Rethinking the unity of science. *International Journal of General Systems, 30*(3), 379–409.

Midgley, G. (Ed.). (2003). *Systems thinking (4 Vols)*. London: SAGE.

Minati, G. (2006). Manifesto: Towards a new generation of systems science societies. *Res-Systemica, 6*(1).

Mingers, J. (1997). Systems typologies in the light of autopoiesis: A reconceptualization of Boulding's hierarchy, and a typology of self-referential systems. *Systems Research and Behavioral Science, 14*(5), 303–313.

Mobus, G. E., & Kalton, M. C. (2014). *Principles of systems science* (2015th ed.). New York: Springer.

Pennock, M. J., & Wade, J. P. (2015). The top 10 illusions of systems engineering: A research agenda. *Procedia Computer Science, 44*, 147–154.

Pickel, A. (2007). Rethinking systems theory. *Philosophy of the Social Sciences, 37*(4), 391–407.

Pouvreau, D. (2011). General systemology as founded and developed by Ludwig von Bertalanffy – An hermeneutical system. In *Paper presented to the conference celebrating Von Bertlanffy's 110th birthday*, Bertalanffy Center for the Study of Systems Science, Vienna 9–10 Nov 2011.

Pouvreau, D. (2013). The project of "general systemology" instigated by Ludwig von Bertalanffy: Genealogy, genesis, reception and advancement. *Kybernetes, 42*(6), 851–868.

Pouvreau, D. (2014). On the history of Ludwig von Bertalanffy's "general systemology", and on its relationship to cybernetics – Part II: Contexts and developments of the systemological hermeneutics instigated by von Bertalanffy. *International Journal of General Systems, 43*(2), 172–245.

Pouvreau, D., & Drack, M. (2007). On the history of Ludwig von Bertalanffy's "General Systemology", and on its relationship to cybernetics, Part 1. *International Journal of General Systems, 36*(3), 281–337.

Ramage, M., & Shipp, K. (2009). *Systems thinkers*. London: Springer.

Rapoport, A. (1968). General system theory. In D. L. Sills (Ed.), *The international encyclopedia of social sciences* (Vol. 15, pp. 452–458). New York: Macmillan & The Free Press.

Rapoport, A. (1976). General systems theory: A bridge between two cultures. Third annual Ludwig von Bertalanffy Memorial Lecture. *Behavioral Science, 21*(4), 228–239.

Rapoport, A. (1986). *General system theory: Essential concepts & applications*. Cambridge, MA: Abacus.

Rousseau, D. (2011). Near-death experiences and the mind-body relationship: A systems-theoretical perspective. *Journal of Near-Death Studies, 29*(3), 399–435.

Rousseau, D. (2014a). A systems model of spirituality. *Zygon: Journal of Religion and Science, 49*(2), 476–508.

Rousseau, D. (2014b). Foundations and a framework for future waves of systemic inquiry. Presented at the 22nd European Meeting on Cybernetics and Systems Research (EMCSR 2014), 2014, Vienna, Austria. In J. M. Wilby, S. Blachfellner, & W. Hofkirchner (Eds.), *EMCSR 2014: Civilisation at the crossroads – response and responsibility of the systems sciences, book of abstracts* (pp. 428–434). Vienna: EMCSR.

Rousseau, D. (2014c). General systems theory: Past, present and potential. Ludwig von Bertalanffy Memorial Lecture. In *58th conference of the international society for the systems sciences, George Washington University*, Washington, DC, July 27 – August 1, 2014. In Learning across the boundaries: Exploring the variety of systemic theory and practice (pp. 35–36). Washington, DC: ISSS.

Rousseau, D. (2014d). Reconciling spirituality with the naturalistic sciences: A systems-philosophical perspective. *Journal for the Study of Spirituality, 4*(2), 174–189.

Rousseau, D. (2014e). Systems philosophy and the unity of knowledge. *Systems Research and Behavioral Science, 31*(2), 146–159.

Rousseau, D. (2015a). General systems theory: Its present and potential [Ludwig von Bertalanffy Memorial Lecture 2014]. *Systems Research and Behavioral Science*, Special Issue: ISSS Yearbook, *32*(5), 522–533.

Rousseau, D. (2015b, May). Systems Philosophy and its relevance to Systems Engineering. Webinar presented as Webinar 76 of the International Council on Systems Engineering.

Rousseau, D. (2016a). *Prospects for the emergence of a general systems worldview.* Contribution to a panel discussion on 'Prospects for a New Systemic Synthesis', together with William (Bill) Schindel, Len Troncale, John Kineman and Jennifer Wilby. Presented at the 60th conference of the International Society for the Systems Sciences (ISSS), Boulder, USA, July 24–30, 2016.

Rousseau, D. (2016b). *Scientific principles for a general theory of whole systems.* Keynote presentation at the 60th annual conference of the International Society for the Systems Sciences (ISSS), Boulder, USA , July 24–30, 2016.

Rousseau, D. (2016c). *Systems science and the quest for a general systems theory.* Research Seminar presented at the Centre for Systems Studies, University of Hull, UK, 19 December 2016.

Rousseau, D. (2017a). *A systems philosophy framework and methodology for discovering and leveraging scientific systems principles.* Presented at the Poster presented at the NSF Engineering and System Design (ESD) and Systems Science (SYS) Programs Workshop 2017: Future directions in engineering design and systems engineering. 20–22 Jan 2017, Georgia Institute of Technology, Atlanta, Georgia, USA. Retrieved from https://drive.google.com/open ?id=0ByXpzF4Gx3NVYkF4X1Aycko0RmM

Rousseau, D. (2017b). *Classifying the Systems Science Body of Knowledge (SSBOK).* Presentation to the Systems Science Working Group (SSWG) Workshop on Classifying the SSBOK. INCOSE International Workshops 2017 (IW'17) – Torrance, CA, USA, 28–31 January 2017.

Rousseau, D. (2017c). *Emerging prospects for establishing systems principles.* Contribution to Panel Discussion on 'Exploring the Frontiers of Systems Science: Establishing a Foundation for Systems Engineering Practice' at the International Symposium 2017 of the International Council on Systems Enineering, Adelaide Australia 17–20 July 2017. Panel moderated by James Martin, other panellists Rick Dove, Anand Kumar, Tim Ferris, Stephen Cook.

Rousseau, D. (2017d). *General systems principles in theory and practice.* Webinar presented to the NASA systems Engineering Research Consortium, NASA Marshall Space Flight Center in Huntsville, AL, United States, on the 15th Nov 2017.

Rousseau, D. (2017e). *Scientific systems principles.* Presentation to the systems philosophy workshop on principles, patterns, isomorphic processes and schemas. INCOSE International Workshops 2017 (IW'17) – Torrance, CA USA, 28–31 January 2017.

Rousseau, D. (2017f). Strategies for discovering scientific systems principles. *Systems Research and Behavioral Science, 34*(5), 527–536.

Rousseau, D. (2017g). *Systems engineering and the quest for scientific systems principles.* Webinar Presented to the Members of the Systems Engineering Community of Practice in Modern Technology Solutions Inc., 9th of June 2017.

Rousseau, D. (2017h). *Systems engineering and the quest for a general science of systems.* Research Seminar Presented at the School of Mechanical, Industrial and manufacturing engineering, Oregon State University, Corvallis OR, 25 Jan 2017.

Rousseau, D. (2017i). Systems research and the quest for scientific systems principles. *Systems, 5*(2), 25.

Rousseau, D. (2017j). *Systems science for systems engineering.* Plenary presentation to the INCOSE International Workshops 2017 (IW'17) – Torrance, CA, USA, 28–31 January 2017.

Rousseau, D. (2017k). *Three general systems principles and their derivation: Insights from the philosophy of science applied to systems concepts.* Proceedings of the 15th annual conference on systems engineering research – Disciplinary convergence : Implications for systems engineering research (23–25 March 2017).

Rousseau, D. (2017l). *Towards a theoretical foundation for SE practice: Challenges, opportunities and progress in the quest for scientific systems principles.* Webinar presented at the annual meeting of the Crossroads of America Chapter of INCOSE, 17th of October 2017.

Rousseau, D. (2017m). *Towards a theoretical foundation for SE practice: Challenges, opportunities and progress in the quest for scientific systems principles.* Webinar presented to the members of the systems engineering community of practice in Roll-Royce, 7th of November 2017.

Rousseau, D. (2018a). *General principles for a science of systems*. Presentation to the Model Based Systems Engineering (MBSE) Working Group during the International Workshops 2018 (IW18) of the International Council on Systems Engineering, held in Jacksonville, Florida USA, 20–23 Jan 2018.

Rousseau, D. (2018b). *Scientific systems principles: A culture change for most systems engineers*. Plenary presentation to the International Workshops 2018 (IW18) of the International Council on Systems Engineering, held in Jacksonville, Florida USA, 20–23 Jan 2018.

Rousseau, D. (2018c). *Systems principles and worldviews*. Workshop presented in the Systems Science Working Group (SSWG) in the International Workshops 2018 (IW18) of the International Council on Systems Engineering, held in Jacksonville, Florida USA, 20–23 Jan 2018.

Rousseau, D. (2018d). *The evolution, variety and uses of scientific systems principles*. Workshop presented in the Systems Science Working Group (SSWG) at the International Workshops 2018 (IW18) of the International Council on Systems Engineering, held in Jacksonville, Florida USA, 20–23 Jan 2018.

Rousseau, D. (2018e). *Three general scientific systems principles for a science of systems*. Workshop presented in the Systems Science Working Group (SSWG) in the International Workshops 2018 (IW18) of the International Council on Systems Engineering, held in Jacksonville, Florida USA, 20–23 Jan 2018.

Rousseau, D. (2018f). Three general systems principles and their derivation: Insights from the philosophy of science applied to systems concepts. In A. M. Madni, B. Boehm, R. G. Ghanem, D. Erwin, & M. J. Wheaton (Eds.), *Disciplinary convergence in systems engineering research* (pp. 665–681). New York: Springer.

Rousseau, D., Blachfellner, S., Wilby, J. M., & Billingham, J. (2016). A Research Agenda for General Systems Transdisciplinarity (GSTD). *Systema, Special Issue – General Systems Transdisciplinarity, 4*(1), 93–103.

Rousseau, D., Billingham, J., Wilby, J. M., & Blachfellner, S. (2016a). The value of general systemology for translational science. In *Presentation at the 9th international systems sciences symposium: Translational systems science*. Organized by the Center for Agent-based Social Systems Sciences, Tokyo Institute of Technology and held on the 13th of March 2016 in the AV Lecture Theatre, West Building 9, Ookayama Campus, Tokyo Institute of Technology.

Rousseau, D., Billingham, J., Wilby, J. M., & Blachfellner, S. (2016b). In search of general systems theory. *Systema, Special Issue – General Systems Transdisciplinarity, 4*(1), 76–92.

Rousseau, D., Billingham, J., Wilby, J. M., & Blachfellner, S. (2016c). The synergy between general systems theory and the general systems worldview. *Systema, Special Issue – General Systems Transdisciplinarity, 4*(1), 61–75.

Rousseau, D., Calvo-Amodio, J., & Barca, R. (2017). *Systems science: Principles for supporting the evolution of systems*. Workshop held at the annual conference of the American Society for Engineering Management held October 18th – 21st, 2017 in Huntsville, Alabama, USA.

Rousseau, D., & Smith, G. (2017). *Systems science principles and their applications in systems engineering projects*. Presented at the INCOSE EMEA Biennial Workshop held 19–21 September 2017 in Mannheim, Germany.

Rousseau, D., & Wilby, J. M. (2014). Moving from disciplinarity to transdisciplinarity in the service of thrivable systems. *Systems Research and Behavioral Science, 31*(5), 666–677.

Rousseau, D., Wilby, J. M., Billingham, J., & Blachfellner, S. (2015a). In Search of General Systems Transdisciplinarity. In *Presented at the International Workshop of the Systems Science Working Group (SysSciWG) of the International Council on Systems Engineering (INCOSE)*, in Torrance, Los Angeles, 24–27 Jan 2015.

Rousseau, D., Wilby, J. M., Billingham, J., & Blachfellner, S. (2015b). Manifesto for General Systems Transdisciplinarity (GSTD). In *Plenary Presentation at the 59th Conference of the International Society for the Systems Sciences (ISSS)*, Berlin, Germany, 4 August 2015.

Rousseau, D., Wilby, J. M., Billingham, J., & Blachfellner, S. (2015c). Systems Philosophy and its relevance to Systems Engineering, Workshop held on 11 July 2015 at the International

Symposium of the International Council on Systems Engineering (INCOSE) in Seattle, Washington, USA.

Rousseau, D., Wilby, J. M., Billingham, J., & Blachfellner, S. (2015d). Systems Philosophy and its relevance to Systems Engineering, Workshop held on Sunday 2 August 2015 – 9am to 5pm, Scandic Hotel Potsdamer Platz, Berlin, German, under the auspices of the the International Society for the Systems Sciences.

Rousseau, D., Wilby, J. M., Billingham, J., & Blachfellner, S. (2016a). A typology for the systems field. *Systema, Special Issue – General Systems Transdisciplinarity, 4*(1), 15–47.

Rousseau, D., Wilby, J. M., Billingham, J., & Blachfellner, S. (2016b). Manifesto for general systems transdisciplinarity. *Systema, Special Issue – General Systems Transdisciplinarity, 4*(1), 4–14.

Rousseau, D., Wilby, J. M., Billingham, J., & Blachfellner, S. (2016c). The scope and range of general systems transdisciplinarity. *Systema, Special Issue – General Systems Transdisciplinarity, 4*(1), 48–60.

Senge, P. M. (1990). *The fifth discipline: The art and practice of the learning organization.* London: Random House.

Sillitto, H. (2014). *Architecting systems. Concepts, principles and practice.* London: College Publications.

Simon, H. A. (1962). The architecture of complexity. *Proceedings of the American Philosophical Society, 106*(6), 467–482.

Sirgy, J. M. (1988). Strategies for developing general systems theories. *Behavioral Science, 33*(1), 25–37.

Skyttner, L. (2006). *General systems theory: Problems, perspectives, practice* (2nd ed.). Hackensack, NJ: World Scientific Publishing Co..

Soban, D. S., Price, M. A., & Hollingsworth, P. (2012). Defining a research agenda in Value Driven Design: Questions that need to be asked. *Journal of Aerospace Operations, 1*(4), 329–342.

Stokols, D. (2006). Toward a science of transdisciplinary action research. *American Journal of Community Psychology, 38*(1–2), 63–77.

Troncale, L. R. (1978). Linkage propositions between fifty principal systems concepts. In G. J. Klir (Ed.), *Applied general systems research* (pp. 29–52). New York: Plenum Press.

Troncale, L. R. (1984). What would a general systems theory look like if I Bumped Into It? *General Systems Bulletin, 14*(3), 7–10.

Troncale, L. R. (1985). The future of general systems research: Obstacles, potentials, case studies. *Systems Research, 2*(1), 43–84.

Troncale, L. R. (1986). Knowing natural systems enables better design of man-made systems: The linkage proposition model. In R. Trappl (Ed.), *Power, autonomy, Utopia* (pp. 43–80). New York: Plenum.

Troncale, L. R. (1988). The systems sciences: What are they? are they one, or many? *European Journal of Operational Research, 37*(1), 8–33.

Troncale, L. R. (2001). The future of the natural systems sciences. In G. Ragsdell & J. Wilby (Eds.), *Understanding complexity* (pp. 219–237). New York: Springer.

Troncale, L. R. (2006). Towards a science of systems. *Systems Research and Behavioral Science, 23*(3), 301–321.

Troncale, L. R. (2009). Revisited: The future of general systems research: Update on obstacles, potentials, case studies. *Systems Research and Behavioral Science, 26*(5), 553–561.

Troncale, L. R. (2017). *Overview of systems processes theory (SPT) & Spinoffs: Comparing its foundations & utility.* Presented at the International Workshop of the Systems Science Working Group (SysSciWG) of the International Council on Systems Engineering (INCOSE), in Torrance, Los Angeles, 28–31 January 2017.

van Gigch, J. P., & Kramer, N. (1981). A taxonomy of systems science. *International Journal of Man-Machine Studies, 14*(2), 179–191.

von Bertalanffy, L. (1932). Allgemeine Theorie, Physikochemie, Aufbau und Entwicklung des Organismus (Theoretische Biologie – Band I):, Berlin, Gebrüder Borntraeger. 1932: Gebrüder Borntraeger.

von Bertalanffy, L. (1950a). An outline of general system theory. *British Journal for the Philosophy of Science, 1*(2), 134–165.

von Bertalanffy, L. (1950b). The theory of open systems in physics and biology. *Science, 111*(2872), 23–29.

von Bertalanffy, L. (1955). General systems theory. *Main Currents in Modern Thought, 11*, 75–83.

von Bertalanffy, L. (1956). General system theory. General systems [Article Reprinted in Midgley, G. (Ed.), (2003) *Systems thinking* (Vol 1 pp 36–51). London: Sage. Page Number References in the Text Refer to the Reprint.], 1, 1–10.

von Bertalanffy, L. (1968). *Organismic psychology and systems theory*. Worcester, MA: Clark University Press.

von Bertalanffy, L. (1969). *General system theory: Foundations, development, applications*. New York: Braziller.

von Bertalanffy, L. (1972). The history and status of general systems theory. *Academy of Management Journal, 15*(4), 407–426.

von Bertalanffy, L. (1975). *Perspectives on general system theory*. New York: Braziller.

von Bertalanffy, L. (1976). *General system theory: Foundations, development, applications* (Rev. ed.). New York: Braziller.

Warfield, J. N. (1990). 'The Great University', a seminar series held at George Mason University, 1989–90 academic year to explore the concept of what would constitute a great university, and how such an institution might be established.

Warfield, J. N. (2003). A proposal for systems science. *Systems Research and Behavioral Science, 20*(6), 507–520.

Whitney, K., Bradley, J. M., Baugh, D. E., & Jr, C. W. C. (2015). Systems theory as a foundation for governance of complex systems. *International Journal of System of Systems Engineering, 6*(1–2), 15–32.

Wiener, N. (1961). *Cybernetics: Control and communication in the animal and the machine* (2nd Rev. ed.). Cambridge, MA: MIT Press.

Wilby, J. M. (2006). An essay on Kenneth E. Boulding's general systems theory: The skeleton of science. *Systems Research and Behavioral Science, 23*(5), 695–699.

Wilby, J. M. (2011). A new framework for viewing the philosophy, principles and practice of systems science. *Systems Research and Behavioral Science, 28*(5), 437–442.

Wilby, J. M. (2014). Boulding's social science gravimeter: Can hierarchical systems theory contribute to its development? In *Proceedings of the 22nd European Meeting on Cybernetics and Systems Research (EMCSR 2014), 2014, Vienna, Austria*, 443–446.

Wilby, J. M. (2016). *Enhancing systemic methodologies*. Contribution to a panel discussion on 'Prospects for a New Systemic Synthesis', together with William (Bill) Schindel, Len Troncale, John Kineman and David Rousseau. Presented at the 60th Conference of the International Society for the Systems Sciences (ISSS), Boulder, USA, July 24–30, 2016.

Wilby, J. M., Macaulay, L., & Theodoulidis, B. (2011). Intentionally holistic knowledge intensive service systems. *International Journal of Services Technology and Management, 16*(2), 126–140.

Wilby, J. M., Rousseau, D., Midgley, G., Drack, M., Billingham, J., & Zimmermann, R. (2015). Philosophical foundations for the modern systems movement. In M. Edson, G. Metcalf, G. Chroust, N. Nguyen, & S. Blachfellner (Eds.), Systems Thinking: New Directions in Theory, Practice and Application, *Proceedings of the 17th conversation of the international federation for systems research, St. Magdalena, Linz, Austria, 27 April – 2 May 2014* (pp. 32–42). Linz, Austria: SEA-Publications, Johannes Kepler University.

Wolkenhauer, O., & Green, S. (2013). The search for organizing principles as a cure against reductionism in systems medicine. *FEBS Journal, 280*(23), 5938–5948.

Wright, C. (2012). *Multiple systems thinking methods for resilience research (MPhil)*. Cardiff University.

Chapter 2
A Disciplinary Field Model for Systemology

Abstract The field of systems is still a nascent academic discipline, with a high degree of fragmentation, no common perspective on the disciplinary structure of the systems domain, and many ambiguities in its use of the term "General Systems Theory". In this chapter we develop a generic model for the structure of a discipline (of any kind) and of disciplinary *fields* of all kinds, and use this to develop a Typology for the domain of systems.

We identify the domain of systems as a transdisciplinary field, and reiterate proposals to call it "Systemology" and its unifying theory GST* (pronounced "G-S-T-star"). We propose names for other major components of the field, and present a tentative map of the systems field, highlighting key gaps and shortcomings. We argue that such a model of the systems field can be helpful for guiding the development of Systemology into a fully-fledged academic field, and for understanding the relationships between Systemology as a transdisciplinary field and the specialized disciplines with which it is engaged.

Keywords Systemology · Transdisciplinarity · General systemology · General systems transdisciplinarity · GSTD · Systems philosophy · General systems theory · GST

2.1 Introduction

As discussed in Chap. 1, prior to the GSTD project there were no available models representing the systems field *as a disciplinary field*, and this has been a significant impediment to establishing a vision of the scope and potential of GSTD, and to prioritizing efforts central to the advancement of GSTD. In the present chapter we present the outcome of a GSTD project aimed at developing such a model and showing how it might be applied.

© David Rousseau 2018
D. Rousseau et al., *General Systemology*, Translational Systems Sciences 13,
https://doi.org/10.1007/978-981-10-0892-4_2

2.2   A Generic Model of a Discipline

In our view, the most urgent issue to be resolved in addressing the academic challenges of the systems domain was to resolve the basic terminological ambiguities in
referring to the field and its components, so that a clear strategy can be formulated
for dealing with the field's scientific challenges. We proposed that this can be
achieved in a systematic way by mapping the components of the field onto the structure of an academic discipline. This strategy was suggested to us by noting that the
various meanings assigned to the term "GST", as discussed earlier, are all components of an academic discipline. We considered that using the structure of a discipline to order the body of knowledge about systems might make it possible to name
the components of this body in an uncontrived manner.

2.2.1   A Systems Model of Discipline

The main objective of the present chapter is to suggest such a structure and nomenclature as a starting point for disambiguating the use of the term "GST", in order to
provide appropriate focus for the development of a general theory of systems that
could underpin the unification and advancement of the systems field. A secondary
benefit of this work is that the disciplinary model developed for the systems field
can be used to support a discussion about the scope, range, completeness and potential of the systems field. It is our hope that the development of a general theory of
systems and the application of this disciplinary model can help to establish the field
of systems as a unified and significant field of academic endeavour.

An immediate challenge for the project to develop a map of the systems field was
that there no clearly established models for the structure of a discipline. In collo-
quial terms a reference to a discipline, for example "Genetics" or "Psychiatry" is
usually meant to designate either a body of knowledge about some subject matter or
a form of action centred on an area of interest or concern. Russ Ackoff once made
such a distinction in the context of systems science, suggesting that general systems
theorists tend to look at science as a body of facts, laws and theories, while systems
researchers look at science as an activity (Ackoff, 1964).

From a systems perspective we can see that both views are apt but incomplete,
not only because each view de-emphasizes the other while both are important, but
also because neither acknowledges the influence of *attitude* (or more specifically,
worldview). Any disciplinarian's worldview motivates and constrains the focus of
their actions, and determines the meanings they ascribe to their data, theories, methods and outcomes (Sutherland, 1973, p. 121). From this perspective we can see that
a discipline is really a kind of system, comprising *a form of action* conditioned by *a
worldview* and expressing *a body of knowledge* centered on some area of interest.
The evolving body of knowledge belonging to a discipline not only *informs* its
worldview but derives its *meaning* from the discipline's worldview.

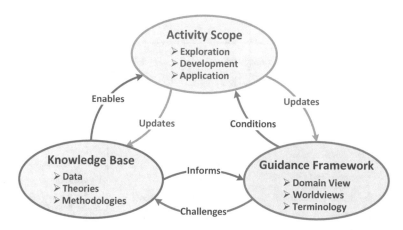

Fig. 2.1 The AKG model of a discipline. (Reproduced from (Rousseau, Wilby, Billingham, & Blachfellner, 2016a), with permission)

In this light we propose that a discipline can be modelled as a system comprising an "activity scope" that is enabled by a "knowledge base" but conditioned by a "guidance framework", as shown in Fig. 2.1. We call this the "Activity-Knowledge-Guidance Model of a Discipline" or "AKG model" for short. Figure 2.1 shows the main elements of a disciplinary system and the ways in which they inter-depend. Each of the main elements has components that are again interdependent but for simplicity these subcomponents are merely listed. These components have internal subdivisions too.

An interesting point highlighted by this model is that the Guidance Framework of a discipline typically involves multiple worldviews. The same subject matter can be studied from different worldviews, and the theories around a given subject can be interpreted differently from different worldview perspectives (Sutherland, 1973, p. 121). Such different approaches to the same subject matter give rise to "disciplinary schools" within a discipline. The schools have the body of knowledge in common, but their different worldviews differentially guide the interpretations and activities of the schools' adherents. For example, within Biology the naturalistic school and the creationist school have different interpretations of the meaning of the theory of evolution, and have different perspectives on the purpose of studying the natural world, and on how knowledge about the natural world may be used. In general, references to a discipline are actually references to the dominant school, and the competing schools are identified by qualifications such as "creationist" or "realist" or "constructivist". In this way, an unqualified reference to a biologist usually refers to a naturalist (rather than for example a creationist), while an unqualified reference to a social scientist usually refers to a constructivist (rather than for example a critical realist).

The model given in Fig. 2.1 does not show the environment within which the system exists and functions, nor does it model the inputs and outputs of the disciplinary system, as these aspects are not crucial for present purposes. However these

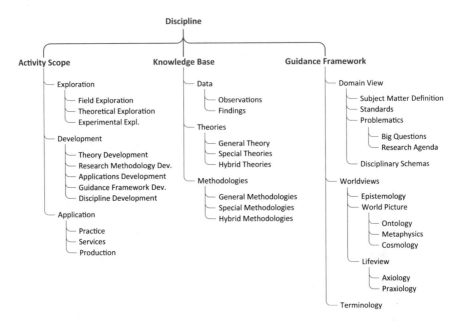

Fig. 2.2 A hierarchical representation of the AKG model of a discipline. (Reproduced from (Rousseau, Wilby, et al., 2016a), with permission)

aspects are important, and we intend to discuss them on another occasion. For present purposes we will focus on the lower-level detail of the AKG model. To expand the AKG model in a manageable way we will from now on show it using a tree structure, as shown in Fig. 2.2. Such a hierarchy preserves containment relationships but unfortunately it obscures the dynamic interactions between the system components. However, it has the important advantage that it can be expanded to show increasing levels of detail as needed.

The structure and subdivisions of Fig. 2.2 broadly follow conventional understandings of the terms used, but some differences necessarily arise because of the attempt to be comprehensive without getting bogged down in pedantry about terms. For this reason, it will be useful to give a brief outline of the conceptual terrain captured by the terms and relationships depicted in Fig. 2.2.

2.2.2   The Disciplinary Activity Scope

The disciplinary activity range describes the actions that represent disciplinarity. These comprise:

1. **Exploration**, being research activities that include:

    (i) **Field Exploration**, research aimed at describing the subject matter in its natural context;

(ii) **Theoretical Exploration**, research aimed at identifying alternative possible interpretations of the field observations and generating hypotheses for testing; and

(iii) **Experimental Exploration**, research aimed at testing hypotheses under partially controlled conditions.

2. **Development**, involving research and reflection towards:

   Theory Development, to update or extend disciplinary theories to accommodate the findings of experimental exploration;

   (i) **Research Methodology Development**, to use the insights from theory development to provide new/improved research methodologies;

   (ii) **Application Development**, to use the findings and insights arising from exploration and theory development to develop new/improved methods for professional practice and physical production, and new/improved designs for products and service systems;

   (iii) **Guidance Framework Development**, to adjust the discipline's guidance framework in the light of the meanings and implications of the findings and insights; and

   (iv) **Discipline Development**, work aimed at sustaining, improving and expanding the discipline as such, for example the development of disciplinary standards for conduct and education, and the development of disciplinary targets and priorities.

3. **Application**, involving using disciplinary knowledge and skills to enable:

   (i) **Professional Practice** that addresses specific problems of individuals by giving advice, taking action or providing support;

   (ii) **Services** provided via service systems that address, for example, general human needs for safety, health, education, dignity; and

   (iii) **Production** of materials, equipment and infrastructure that support individual and social welfare.

2.2.3   The Disciplinary Knowledge Base

The disciplinary knowledge base comprises the key resources that enable disciplinary activity. These comprise:

1. **Data**, consisting of:

   (i) **Observations,** being descriptions of the subject matter as encountered in ordinary contexts. These include descriptions of the subject matter entities in terms of their appearance, structure, behaviour, powers, and functions; and

   (ii) **Findings**, representing the outcomes of experiments and tests under partially controlled conditions.

2. **Theories**, consisting of:

  (i) a **General Theory**, i.e. a theory that applies always and everywhere within the discipline, and is the basis of its scientific unity, for example the Periodic Table of Elements in chemistry and the Theory of Evolution by Natural Selection in Biology[1];

  (ii) Special **Theories**, i.e. theories about subclasses of the subject matter. For example, in Chemistry these include theories about classes of chemicals, for example metals, radioactive isotopes, polymers; and

  (iii) **Hybrid Theories**, i.e. theories that combine special theories with theories from other disciplines when interests overlap. For example, in the case of Chemistry these are hybrid theories such as those of Biochemistry, Geochemistry, Nuclear Chemistry, and Neurochemistry.

3. **Methodologies**, consisting of:

  (i) **General Methodologies**, i.e. disciplinary ways of working that are of general utility across the specializations of the discipline;

  (ii) **Special Methodologies**, i.e. structured ways of tackling specialized kinds of disciplinary problems; and

  (iii) **Hybrid Methodologies**, i.e. structured ways of tackling problems involving multiple disciplines. In substantive cases they become the methodologies of Hybrid Disciplines.

2.2.4 The Disciplinary Guidance Framework

The disciplinary guidance framework provides the context that conditions disciplinary activity, giving direction and focus, and setting boundaries, standards and priorities. More specifically, it involves:

1. A **Domain View**, comprising:

  (i) a **Subject Matter Definition** that specifies the scope and range of the discipline's interests;

  (ii) **Standards** for governing professional conduct and ensuring quality;

  (iii) a **Problematics** comprising:

   • The "**Big Questions**" the discipline seeks to answer;
   • A **Research Agenda** that defines and prioritizes the work of the discipline; and

  (iv) **Disciplinary Schemas** that map the relationships between the components of the discipline.

2. A **Worldview**, comprising:

---

[1] We provide a more extensive discussion of the structure of a general theory in Chap. 5.

(i) an **Epistemology**, that explains what knowledge is, describes what enables, conditions or prevents the acquisition of kinds of knowledge, discusses opportunities for and limits on what we can come to know; and explains how the models and theories of the discipline can be used to acquire knowledge relevant to the purposes of the discipline; and

(ii) a **World Picture** comprising:

- An Ontology, i.e. a theory of what exists most fundamentally, for example "physical atoms", or "God" or "Tao";
- A Metaphysics, i.e. a theory about the nature of what exists and hence what is possible, for example "all changes are proportional to changes elsewhere", or "all events have sufficient reasons", or "all outcomes are due to Divine providence"; and
- A Cosmology (model of the origin, history, organisation and possible futures of the concrete world[2]).

3. A **Lifeview**, comprising:

- an **Axiology** (a value system and theories about the nature of values and how to make value judgements); and
- a **Praxeology** (theory about the nature of action, agency, freedom and responsibility).

4. A **Terminology** that provides the standard terms and coherent concepts needed for model building in the discipline's domain of operation.

2.2.5 Kinds of Disciplines

The AKG model provides a way of distinguishing between a topic, theory or activity, and a complete discipline. A discipline, in this light, is an interconnected system, comprising activities that, under the conditioning influence of a guidance framework, produce outputs that include updating knowledge about a defined subject matter. The term "discipline" so defined is clearly very broad, and hence it can be used to characterise a variety of kinds of disciplines, which we differentiate as follows.

As discussed earlier, theories can be either general, specialized or hybrid theories, and hence the methodologies they enable can be either general, specialized or hybrid methodologies. The general theory that characterizes the subject matter of the discipline applies in and connects the special and hybrid theories/methodologies, and in this sense is a "meta-theory" over the special and hybrid theories/methodologies, thereby forming the basis of the unity of the discipline.

---

[2]Things are "concrete" if they have causal powers; this distinguishes them from *abstract* things, which can also be considered to be "real" in the sense of having existence independently of our imagination (for example numbers) but that do not have causal powers.

As a discipline matures its theories and methodologies become rich and diverse, and this gives rise to sub-disciplines dedicated to refining, extending, promoting and applying the original discipline's individual theories or methodologies. In this way a strong discipline soon becomes a "disciplinary field", divisible into general, special and hybrid *disciplines*. In this case the general theory (meta-theory) of the field becomes a special case of a *transdisciplinary* theory, because it now applies in and connects between the special and hybrid disciplines of that field. In this way the "general discipline" in a field is a "transdiscipline" that applies across the special and hybrid disciplines of the field, and is also the discipline that underpins and develops the scientific unity of the disciplinary field. The disciplines commonly encountered across academic institutions are the most advanced ones, and hence the disciplinary divisions we typically encounter in academia are disciplinary fields.

An interesting observation that follows from looking at disciplines and fields in this way is that there is a meta-theory at the heart of every discipline, and a transdiscipline at the heart of every disciplinary field. The scope of such meta-theories and transdisciplines is however typically limited to the scope of the discipline or field they unify. This represents a special case of transdisciplinarity, different from how it is usually discussed, namely as applying across the major traditional academic divisions we have here identified as fields. However this framing follows directly from the basic meanings of the terms 'transdisciplinarity' and 'discipline'. This does not eliminate or replace the idea of a transdisciplinarity that crosses the boundaries between fields, but it does indicate that there are different kinds of transdisciplinarity which we should be careful to disambiguate. The terms we are proposing in this section will enable us to explore the nature of transdisciplinarity in more detail in Chap. 3.

As noted earlier, disciplines fragment into schools based on differences in worldviews such as Naturalism, Creationism, and Constructivism. However, within a field there are also connections between the schools that share a worldview, so that together they form a community of practice we call a disciplinary "tradition" within the field. A tradition opens up channels of communication and co-operation between schools, via the perspectival unity provided by the common worldview. These channels extend beyond the disciplinary field to also facilitate communication and co-operation with consilient schools in other fields. This is powerful for the schools associated with the dominant tradition in a field, but it can also be a limiting factor by inhibiting exploration of alternative perspectives and reducing sensitivity to the inherent fallibility of human perspectives (Casadevall & Fang, 2015; Kuhn, 1996).

2.2.6 The Structure of a Disciplinary Field

Given the distinctions just enumerated, we can conceptualise the structure of a disciplinary field in terms of its constituent disciplines as illustrated schematically in Fig. 2.3. We call this the "disciplinary spectrum" model of a field. For simplicity

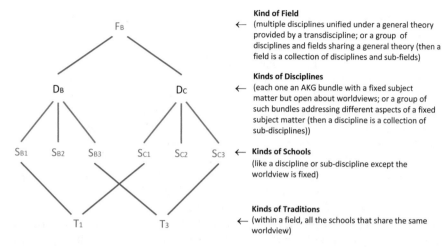

Fig. 2.3 The disciplinary spectrum model of a disciplinary field. (Adapted from (Rousseau, Wilby, et al., 2016a))

only three disciplines are shown in the diagram, but in practice there will typically be many.

A *discipline* can be viewed as something that has the tripartite content structure we elaborated earlier, comprising an activity scope, a body of knowledge, and a guidance framework (the AKG model), and has a fixed subject matter but not a fixed worldview. If the worldview is fixed, then we have a *school* within the discipline. A discipline can be comprised of *sub-disciplines*, each focused on a specific aspect of the disciplinary subject matter. A collection of disciplines unified under a general theory constitutes a *field*, and as the general theory is then transdisciplinary the discipline that provides it a (unifying) *transdiscipline*. Within a field there can be various *traditions*, represented by the schools that share a common worldview.

Every discipline, school and tradition in the field will have the tripartite content structure we elaborated earlier, and we refer to this in Fig. 2.3 as "an AKG bundle'''. The field includes the contents of all its constituent disciplines, and therefore it also has the AKG structure in terms of its contents, as previously shown in Fig. 2.2 (except that the top label "discipline" is replaced by "field").

It should however be noted that the field is more than merely the sum of its constituent disciplines. The field's structure establishes systemic relationships between the constituents that both limits and empowers them, and the whole provides a stronger basis for the development of the constituents by placing them in context relative to other disciplinary fields. The status and strength of the field lends credibility to its constituent disciplines and schools, creating opportunities for funding, recruitment and participation, and providing connections that stimulate theoretical and methodological innovation. On the other hand the field also constrains its components by introducing standards, regulating behaviour, setting priorities, and so on. The field is unified by the general theory that is the same for all the disciplines.

In practice the situation can be even more complicated, and so we have to recognize the existence of fields that have both fields and disciplines as components, in which we can call the component fields "sub-fields" and the overarching field a "super-field". However, for simplicity we will use the term "field" for all these special cases, unless it would be confusing in the discussion context.

For example, we can view science as a field that includes subfields such as physics, chemistry and biology as well disciplines such as philosophy of science. Science (as the study of nature) is unified under a shared theory about the nature of nature as comprehensible and investigable. Biology is a subfield of science that unites the biological disciplines under the theories of evolution and genetics. Biology disciplines such as plant biology have many sub-disciplines studying aspects of plants (for example plant toxicology) or kinds of plants (for example xerophytes). Biology contains multiple schools for example the naturalistic school and the creationist school, and these schools are the biology representatives of the naturalistic and creationist traditions in the field of science.

With these two models (AKG model and spectrum model) in hand it is possible to draw a "map of the territory" for any field from either or both of two complementary perspectives: its disciplinary composition in terms of the spectrum model and the organization of its content in terms of the AKG model.

2.3   Systemology Modelled As a Disciplinary Field

By applying these two models we can now begin to characterise the systems domain in disciplinary terms. To do this, we have to select suitable names for the various elements of the systems discipline.

2.3.1   The Nature of the Systems "Discipline"

In the light of the analysis just given, systems science is a disciplinary field containing the general discipline of general systemology, many specialised systems disciplines (for example Cybernetics, Management Science, and Operational Research), and many hybrid systems disciplines (for example systems biology and systems psychology). These disciplines can all be represented by schools grounded in specific worldviews such as in Scientific Realism or Constructivism.

The disciplinary schools can be grouped into traditions for example the Scientific Realist Tradition, the Constructivist Tradition and the Postmodern Tradition. Each of these traditions span across the divisions into philosophy, science, engineering and practice.

As is the case in general, the systems disciplines typically have dominant schools, for example general systems theorists are typically critical realists, second-order cyberneticians are typically radical constructivists, and management scientists are typically pragmatic pluralists. It is unclear whether there is presently a dominant tradition within the field of systems as a whole.

It should be noted that schools and traditions should be named with caution, because the worldviews they represent are typically complex – for example, a worldview might be realistic about certain things (for example natural systems) and constructivist about others (for example value systems). We have discussed this issue elsewhere (Wilby et al., 2015). A principled way of dealing with this problem is to name a school or tradition after its perspective on its subject matter. This can produce surprising results, for example the presently dominant school in Systems Engineering would then be classified under the tradition of Moderate Constructivism: although most systems engineers are objectivist-realists about the existence of physical entities in the concrete world (including the products produced via Systems Engineering), they typically are (non-radical) constructivists about whether anything is *inherently* a system (Sillitto et al., 2017).

2.3.2 *Naming the Systems Field*

There is no established English term for the field of systems as a whole. Terms such as Systems Thinking, Systems Research, Systems Science, Complexity Science and Systems Theory are widely used, but there terms are perhaps too narrow in their native meanings to be compelling as a globally encompassing reference. There is an apt *German* term for the field, *Systemlehre*, meaning "an organised body of knowledge about systems". Bertalanffy scholars David Pouvreau and Manfred Drack have suggested that an apt translation of the concept *Systemlehre* is provided by the term "Systemology" (Pouvreau & Drack, 2007, pp. 282–283), and in fact "Systemology" was suggested even earlier by Russ Ackoff, saying:

> As the problem complexes with which we concern ourselves increase in complexity, the need for bringing the interdisciplines together increases. What we need may be called *meta-disciplines*, and what they are needed for may be called *systemology* (Ackoff, 1973, p. 669).

The term "Systemology" was used by Jan Kamarýt in his contribution to the *Festschrift* for von Bertalanffy, *Unity Through Diversity* (Kamarýt, 1973, p. 88), but the term seems to have fallen into disuse later on. However "Systemology" *is* an apt term, because etymologically it would literally mean "what could be said about systems", or in practice "the study of systems". We therefore recommend that "Systemology" be revived as the name for the systems field, to encompass the specialized systems disciplines and sub-fields such as systems philosophy, systems science, systems engineering and systems practice.

2.3.3 The General Theory of Systemology

As discussed earlier, the crucial step along the path to becoming an academically viable disciplinary field is the establishment of a unifying theory. In the case of Systemology, this would be a general theory about the kinds, nature and evolution of systems.

Von Bertalanffy had proposed that there exists, in principle, a theory encompassing "the universal principles applying to systems in general" (von Bertalanffy, 1956, p. 37), and this was one of the meanings assigned to his use of the term "GST", sometimes referred to by himself and others as "GST in the narrow sense". This meaning corresponds to the notion of a "general theory" for Systemology, and is the most natural use of the term "GST", unlike the many other uses of the term as discussed in Chap. 1. We would prefer to see the scope of the term "GST" reduced to refer only to this "universal" theory. However, in order to ensure disambiguation from the many other uses of the term "GST" in the literature, we propose that *this* theory be called GST* (pronounced "g-s-t-star").

2.3.4 The Unifying Transdiscipline of Systemology

Apart from proposing a general theory, von Bertalanffy also called for the establishment of a new discipline the subject matter of which is the derivation and formulation of the general systems principles, with a view to putting them to use to empower all the disciplines dealing with systems (von Bertalanffy, 1969, p. 32). Von Bertalanffy referred to this as a "meta-discipline", and originally named it "Allgemeine Systemlehre" in his native German. In English he referred to this first as "GST" and then later on as "GST in the broad sense". This is clearly inadequate and confusing.

"Allgemeine" means "general", and as discussed earlier "Systemlehre" is aptly translated as "Systemology", and therefore an intuitive naming of this transdiscipline reflecting "GST in the narrow sense" would be "General Systemology". Exactly this has in fact been proposed by Pouvreau and Drack (Pouvreau & Drack, 2007, pp. 282–283), and we advocate that this usage be adopted by the systems community.

2.3.5 The Specialized Theories of Systems Science

The "special disciplines" of a field are concerned with developing and applying theories about specialised aspects or elements of the field's subject matter. For systems science (and hence for Systemology) these would be theories about specific kinds of systemic structures or behaviours, for example control theory, network theory, hierarchy theory, automata theory and so on.

Mario Bunge coined the term "Systemics" for this set of special theories (Bunge, 1979, p. 1), and we propose that this usage be standardized in the typology of Systemology. The systems concepts being transdisciplinary, the Systemics are all formal theories,[3] and hence applicable in different kinds of concrete contexts.

For present purposes Bunge's term can usefully be extended in the following way. Strictly speaking, his term refers to "Abstract Theoretical Systemics", in the sense that they are formal theories about special kinds of systemic structures and behaviors; however, note that there are also "Abstract Methodological Systemics", i.e. formal methodologies for analyzing systemic complexity for example in specialised systems disciplines such Systems Dynamics, Systems Analysis, and Operational Research. When the abstract theoretical and methodological Systemics are employed by specialized orthodox disciplines (which have concrete subject matters), this gives rise to hybrid disciplines such as Systems Biology, Systems Geology and Systems Medicine. The theories of the hybrid disciplines can be called "Applied Theoretical Systemics" and their methodologies "Applied Methodological Systemics". The "applied" systemic theories/methodologies differ from the "abstract" ones in that they involve specific ontological commitments, and hence are concrete theories/methodologies rather than formal ones.

Compared to other academic disciplines Systemology is unique in having this structure. In the case of for example Mathematics "pure" Mathematics and Applied Mathematics are both formal disciplines, and in the case of the orthodox sciences a "pure" science and its associated applied science are both concrete disciplines. Systemology however has both formal and concrete dimensions. This observation explains the origins of the long-standing controversy about whether "Systems Science" is (at least in principle) really a science or not – as we can now see, the debate really confronts *Systemology*, and it has elements that are sciences and ones that are not (including formal theories, heuristic disciplines and philosophical disciplines). It also explains why many of the Abstract Theoretical Systemics ("Systemics") are studied in Mathematics departments while the applied ones (specialised and hybrid sciences and systems practices) are not.

An interesting upshot of making this distinction is that the theories and methods of Systems Engineering fall within the "abstract" rather than the "applied" category, because these theories and methods are agnostic about the kind of system involved. This sets Systems Engineering apart from other types of engineering, which are all concrete disciplines.

---

[3] A formal theory is one that makes no ontological commitments, ranging over abstract entities that could be instantiated in many ways. This contrasts with concrete theories, which has specific ontological commitments that are essential for the theory to be valid.

2.3.6   The Transdisciplinary Nature of Systemology

Systemology is an unusual disciplinary field because its core concept, "system" is a transdisciplinary one. From the systems perspective one could characterise all the orthodox disciplines as studying specific kinds of systems, and hence the concepts, principles and models involved in characterizing aspects of systemicity (for example feedbacks and hierarchies) can be applied across the spectrum of orthodox disciplines. Consequently, the special theories, methodologies and disciplines of Systemology are *all* transdisciplinary theories, methods and disciplines.

This sets Systemology apart from orthodox disciplinary fields, because orthodox fields have only one transdiscipline each, namely the one developing the general theory that unites the field.

However, it should be noted that despite containing many transdisciplines Systemology has only one transdiscipline responsible for developing its unifying theory (General Systemology). The other transdisciplines do not serve to unify Systemology, nor do they serve to unify, into fields, the orthodox disciplines they range over. Chapter 3 provides a detailed discussion of the scope and range of General Systemology *as a transdiscipline*.

2.3.7   The Traditions and Schools of the Systems Field

As a field the structure of Systemology is complex, comprising many disciplines each represented by multiple schools associated via multiple traditions. The most developed tradition in Systemology is the one associated with Broad Naturalism and Moderate Scientific Realism. The disciplinary schools in this tradition originated in the "first wave" of systemic inquiry in the 1950s (Midgley, 2000, pp. 187–211), which is particularly associated with physical systems and industrial applications. This tradition is especially associated with von Bertalanffy and the search for a General System Theory, and with the rise of (first-order) Cybernetics and Systems Engineering. Although this is a Naturalistic systems tradition it should be carefully noted that in the systems community this tradition is not essentially physicalistic nor reductionistic. Von Bertalanffy and his followers were strongly opposed to the positivism and behaviourism that were features of the physical sciences and technological industries of his time, and saw the systems approach as preserving the humanism that physicalism and classical reductionism tended to devalue.

The next strongest tradition is associated with Constructivism, and originated in the "second wave" of systemic inquiry in the 1970s; it is now largely associated with organizational design and management science (for example the "Soft Systems Methodology" of Peter Checkland, and the "Second-order Cybernetics" of Gregory Bateson and Heinz von Foerster).

There are many other, smaller groups of schools, largely originating onwards from the 1980s, and representing a diversity of traditions grounded in philosophical positions such as liberalism, holism, postmodernism, and pragmatic pluralism.

2.3.8 The Naturalistic Tradition in Systemology

Von Bertalanffy called his own worldview "Perspectivism" (von Bertalanffy, 1955), and meant by this a view that was Naturalistic but moderated by several supplementary views and reservations, so as to be intermediate between Scientific Realism (or better, Objectivist Realism) and Social Constructivism. His view represented a form of Realism in that he accepted the existence of a universe independent of the observer, and a form of Naturalism in that he accepted that the scientific method can reveal aspects of reality's nature. However his view was moderate in that he accepted that science is limited in what it can reveal, and that the making of observations and the building of models and theories are conditioned both by agents' cognitive capacities and the purposes that agents have in mind.[4] In this light we can see that von Bertalanffy can be identified with a school in General Systemology that falls within a systems tradition that we can call "Naturalistic Systemology". As we argue in Chap. 4, there is a synergistic relationship between GST* and the worldview of Naturalistic Systemology, in that this worldview opens up routes to the discovery of general systems principles, and that these principles enhance the power of this worldview to guide us to the discovery of further general systems principles. In this light we have adopted the traditional naming of this worldview as the "General Systems Worldview (GSW)" (Elohim, 2000; von Bertalanffy, 1934). Our own position falls within the (broad) naturalistic school of General Systemology, but we do recognise that once a GST* exists it could be interpreted under a variety of onto-epistemic positions, so it is by no means suggested that the GSW has a privileged status within Systemology.

---

[4]The term "Scientific Perspectivism" is now becoming established as the contemporary label for this nexus of philosophical commitments (Callebaut 2012; Giere 2006; Jaeger, Laubichler, & Callebaut 2015). Extensive treatments are given in Wimsatt (2007) (who calls his view "Multiperspectival Realism") and van Fraassen (2008) (who calls his view "Constructive Empiricism"). Werner Callebaut argues that Scientific Perspectivism "provides the best resources currently at our disposal to tackle many of the outstanding philosophical issues implied in the modelling of complex, multilevel/multiscale phenomena" (Callebaut 2012, p. 75). The term "Perspectivism" was coined in the mid-1880s by Friedrich Nietzsche in *The Will to Power* (Nietszche 1964, § 481). Although Nietzsche's work in this area informed the development of Scientific Perspectivism, his views were more radical than those of Scientific Perspectivism, and can be viewed as foundational to Postmodernism and Constructivism.

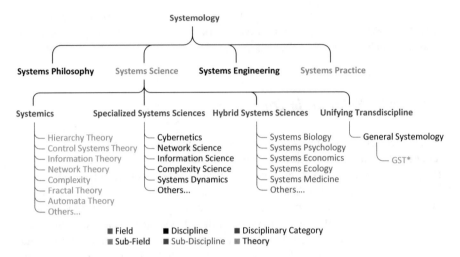

Fig. 2.4 A disciplinary spectrum typology for systemology

2.3.9   A Typology for Systemology

With these clarifications in hand we can now present a typology for Systemology from two perspectives, one showing the disciplinary structure of Systemology (a disciplinary spectrum model of Systemology), as illustrated in Fig. 2.4, and one showing how its content is organised (a hierarchical AKG model of Systemology) as illustrated in Fig. 2.5. For simplicity Fig. 2.4 shows only the upper level of the spectrum, leaving out schools and traditions. In Fig. 2.4 we use multiple colours, and in Fig. 2.5 we use red to highlight the presence and context of the special terms we advocated in this chapter.

In the AKG map shown in Fig. 2.5 we have focused on the Knowledge Base of Systemology. The process of drawing the AKG map showed that Systemology is rich in methodologies (many hundreds) and relatively rich in special theories and hybrid theories (dozens), but poor in material relevant to GST*.

Any discipline within systemology can be analyzed in detail using this structure, as we do in the next section for General Systemology, whose developmental backlog was identified at the outset as one of the root causes of the challenges facing Systemology.

2.4   Assessment of the Developmental Status of General Systemology

In the light of the typological structure developed here we can now undertake a more principled review of the developmental status of General Systemology, as follows.

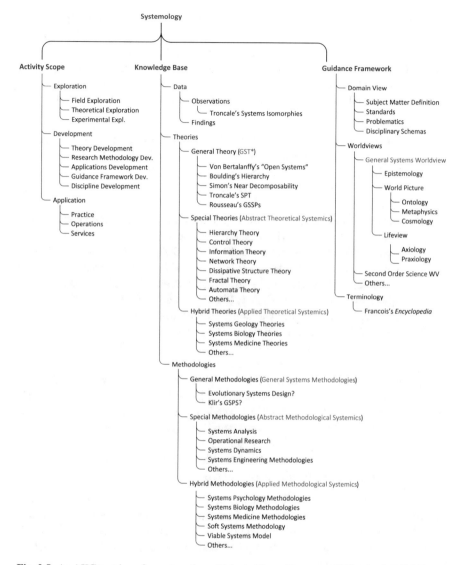

Fig. 2.5 An AKG typology for systemology. (Adapted from (Rousseau, Wilby, et al., 2016a))

1. **Activity Scope**: At the moment we have no established GST*, and hence no GSTD as such, although some researchers are working towards developing and establishing it (Rousseau, Blachfellner, Wilby, & Billingham, 2016; Rousseau, Wilby, Billingham, & Blachfellner, 2015a, 2015b, 2016b).
2. **Knowledge Base**: As yet we have no general theory of systems, but we have interesting and useful components to build on, including von Bertalanffy's proposed general systems principle – that hat there are no closed systems in nature

(von Bertalanffy, 1950), Herbert Simon's 'principle' that all complex systems are near-decomposable hierarchies (Simon, 1962), Ross Ashby's "Law of Requisite Variety" (Ashby, 1956), Len Troncale's "Systems Process Theory" (SPT) (Friendshuh & Troncale, 2012; Troncale, 1978), Rousseau's three general scientific systems principles (discussed in Chap. 6) and, on the methodology side, George Klir's "General Systems Problem Solver" (Klir, 1985, 2001).

3. Guidance Framework:

(i) **General Systems Domain View**: The potential scope and value of General Systemology have been widely discussed (Gray & Rizzo, 1973; Hammond, 2003; Laszlo, 1972b; von Bertalanffy, 1969, 1975), but these presentations were often of wider scope due to the ambiguity of the historical term "GST";

(ii) **General Systems Worldview**: We have no comprehensive synthesis yet, although we have early candidate models in (Laszlo, 1972a, 1972c; Sutherland, 1973) and other works we can draw on. For example, we have works in the areas of **Ontology** (Bunge, 1977), **Metaphysics** (see e.g. the principles listed in Bunge, 1979, 1981; Rousseau, 2015; Sutherland, 1973), **Epistemology** (von Bertalanffy, 1955), **Axiology** (Hammond, 2005; Parra-Luna, 2001, 2008) and **Praxiology** (Kotarbinski, 1965);

(iii) **General Systems Terminology**: Despite the clarifications in the present chapter terminology remains a problematic issue for General Systemology as indeed it does for Systemology as a whole. Charles Francois' *Encyclopedia* (Francois, 2004) is now more than a decade old, and serves more to display the lack of coherence in the field's terminology than to set linguistic standards. Several authors have developed terminologies of their own, to compensate for this problem (Heylighen, 2012, pp. 21–33; Kramer & De Smit, 1977, pp. 11–46; Schindel, 2013; Schoderbek, Schoderbek, & Kefalas, 1990, pp. 13–68; Wimsatt, 2007, pp. 353–360), but there is no programme in hand to unify these efforts.

The incomplete state of GST* and the GSW is a serious impediment to the maturation of Systemology as an academic field, but in the light of the AKG Typology we can see where the key gaps are, and from this develop a focused plan for development. We present a Research Agenda addressing this elsewhere (Rousseau, Blachfellner, et al., 2016).

On the basis of a scientific general systems theory we could reconcile or refine the various definitions we have today of what "system" entails, and develop theoretical foundations for the methodologies that lack them, thus setting them on the path to increasing effectiveness and expanding scope.

GST* would not only provide a scientific unification of the field and extend existing powers, but moreover a strong general theory would open up routes to discovering new abstract Systemics (both theoretical and methodological), and together with a developed GSW would open up new opportunities in exploratory science (Rousseau, Billingham, Wilby, & Blachfellner, 2016; Rousseau et al., 2015b; Rousseau, Wilby, Billingham, & Blachfellner, 2016c). Such advances would contribute in important ways to systemology becoming established as an academic field in its own right.

2.5 Summary

In this Chapter we developed a generic model for the structure of a discipline and of a disciplinary field, and used this to develop a Typology for the domain of systems.

In order to do this we introduce a generic systemic model of a discipline in terms of the interactions between a discipline's activity scope, knowledge base and guidance framework (the "AKG model") and the structure of a disciplinary field in terms of a spectrum of fields, disciplines, schools and traditions (the "disciplinary spectrum model").

Using these models, we developed a typology by:

 (i) identifying the domain of systems as a disciplinary field, and advocating it be named "Systemology";
(ii) identifying the unifying theory of the field as von Bertalanffy's "GST in the narrow sense" and naming it GST* (pronounced "G-S-T-star");
(iii) identifying the transdiscipline GST* would ground as von Bertalanffy's "GST in the broad sense", and adopting "General Systemology" as the name of this transdiscipline; and
(iv) identifying the special theories of the field as corresponding to Bunge's use of the term "Systemics", and correspondingly introducing the class-names "Abstract Theoretical Systemics" and "Applied Theoretical Systemics" and the methodological correspondences in "Abstract Methodological Systemics" and "Applied Methodological Systemics".

We have used the models and naming conventions developed in this Chapter to sketch a preliminary map of the 'systems territory' conceived as a disciplinary field, and explored how to use it to assess and discuss the structure and completeness of Systemology and its components in a non-ambiguous way, and to place the work that is being done to complete or improve systemological components in their proper context. We are hopeful that will lead to further constructive discussions about the nature, structure and completeness of the field of systemology.

Moreover, we have tried to show that the lack of a developed general theory of systems (GST*) is at the root of the fragmentation and limited influence of the systems field, and that progress with such a theory will be key for establishing Systemology academically and enhancing its impact.

We believe that these concepts, models and views will be helpful in formulating agendas and strategies for developing Systemology into an established and valued academic discipline. In the chapters to follow we present further, more detailed reflections on the nature of the theories and worldviews needed to advance this task, and then outline initial scientific progress made in applying these ideas to develop general scientific systems principles.

References

Ackoff, R. L. (1964). General system theory and systems research: Contrasting conceptions of systems science'. *Views on General Systems Theory: Proceedings of the Second Systems Symposium at Case Institute of* Technology, 51–60.

Ackoff, R. L. (1973). Science in the systems age: Beyond IE, OR, and MS. *Operations Research, 21*(3), 661–671.

Ashby, W. R. (1956). *An introduction to cybernetics*. London: Chapman & Hall.

Bunge, M. (1977). *Ontology I: The furniture of the world*. Boston, MA: Reidel.

Bunge, M. (1979). *Ontology II: A world of systems*. Dordrecht, The Netherlands: Reidel.

Bunge, M. (1981). *Scientific materialism*. Dordrecht, The Netherlands: Reidel.

Callebaut, W. (2012). Scientific perspectivism: A philosopher of science's response to the challenge of big data biology. *Studies in History and Philosophy of Science Part C: Studies in History and Philosophy of Biological and Biomedical Sciences, 43*(1), 69–80.

Casadevall, A., & Fang, F. C. (2015). Field science—The nature and utility of scientific fields. *MBio, 6*(5), 1–4.

Elohim, J.-L. (2000). *A general systems WELTANSCHAUUNG (Worldview)*. Retrieved January 20, 2016, from http://www.isss.org/weltansc.htm

Francois, C. (Ed.). (2004). *International encyclopedia of systems and cybernetics*. Munich, Germany: Saur Verlag.

Friendshuh, L., & Troncale, L. R. (2012). Identifying fundamental systems processes for a general theory of systems. *Proceedings of the 56th Annual Conference, International Society for the Systems Sciences (ISSS)*, July 15–20, San Jose State University, 23 pp.

Giere, R. N. (2006). *Scientific Perspectivism*. Chicago: University Of Chicago Press.

Gray, W., & Rizzo, N. D. (Eds.). (1973). *Unity through diversity*. New York: Gordon & Breach.

Hammond, D. (2003). *The science of synthesis: Exploring the social implications of general systems theory*. Boulder, CO: University Press of Colorado.

Hammond, D. (2005). Philosophical and ethical foundations of systems thinking. *TripleC: Communication, Capitalism & Critique. Open Access Journal for a Global Sustainable Information Society, 3*(2), 20–27.

Heylighen, F. (2012). Self-organization of complex, intelligent systems: An action ontology for transdisciplinary integration. *Integral Review*. Retrieved from http://pespmc1.vub.ac.be/papers/ECCO-paradigm.pdf

Jaeger, J., Laubichler, M., & Callebaut, W. (2015). The comet cometh: Evolving developmental systems. *Biological Theory, 10*(1), 36–49.

Kamarýt, J. (1973). From science to metascience and philosophy. In W. Gray & N. D. Rizzo (Eds.), *Unity through diversity* (pp. 75–100). New York: Gordon & Breach.

Klir, G. J. (1985). *Architecture of systems problem solving*. New York: Plenum Press.

Klir, G. J. (2001). *Facets of systems science* (Softcover reprint of the original 2nd ed. 2001st edition). Boston: Springer.

Kotarbinski, T. (1965). *Praxeology: An introduction to the sciences of effective action*. Warsaw, Poland: PWN.

Kramer, N. J. T. A., & De Smit, J. (1977). *Systems thinking: Concepts and notions*. Leiden, The Netherlands: Martinus Nijhoff.

Kuhn, T. (1996). *The structure of scientific revolutions* (3rd ed. (original 1962)). Chicago: University of Chicago Press.

Laszlo, E. (1972a). *Introduction to systems philosophy: Toward a new paradigm of contemporary thought*. New York: Gordon & Breach.

Laszlo, E. (Ed.). (1972b). *The relevance of general systems theory*. New York: George Braziller.

Laszlo, E. (1972c). *The systems view of the world: The natural philosophy of the new developments in the sciences*. New York: George Braziller.

Midgley, G. (2000). *Systemic intervention: Philosophy, methodology, and practice*. New York: Kluwer.

Nietszche, F. (1964). The will to power. In O. Levy (Ed.), (A. M. Ludovici, Trans.). *The complete works of Friederich Nietzsche* (Vol. 15). New York: Russell & Russell.

Parra-Luna, F. (2001). An axiological systems theory: Some basic hypotheses. *Systems Research and Behavioral Science, 18*(6), 479–503.

Parra-Luna, F. (2008). Axiological systems theory: A general model of society. *TripleC: Communication, Capitalism & Critique. Open Access Journal for a Global Sustainable Information Society, 6*(1), 1–23.

Pouvreau, D., & Drack, M. (2007). On the history of Ludwig von Bertalanffy's "General Systemology", and on its relationship to cybernetics, Part 1. *International Journal of General Systems, 36*(3), 281–337.

Rousseau, D. (2015). General systems theory: Its present and potential [Ludwig von Bertalanffy Memorial Lecture 2014]. *Systems Research and Behavioral Science,* Special Issue: ISSS Yearbook, *32*(5), 522–533.

Rousseau, D., Billingham, J., Wilby, J. M., & Blachfellner, S. (2016). The synergy between general systems theory and the general systems worldview. *Systema, Special Issue – General Systems Transdisciplinarity, 4*(1), 61–75.

Rousseau, D., Blachfellner, S., Wilby, J. M., & Billingham, J. (2016). A research agenda for General Systems Transdisciplinarity (GSTD). *Systema, Special Issue – General Systems Transdisciplinarity, 4*(1), 93–103.

Rousseau, D., Wilby, J. M., Billingham, J., & Blachfellner, S. (2015a). In Search of general systems Transdisciplinarity. Presented at the International Workshop of the Systems Science Working Group (SysSciWG) of the International Council on Systems Engineering (INCOSE), in Torrance, Los Angeles, 24–27 Jan 2015.

Rousseau, D., Wilby, J. M., Billingham, J., & Blachfellner, S. (2015b). Manifesto for General Systems Transdisciplinarity (GSTD), Plenary Presentation at the 59th Conference of the International Society for the Systems Sciences (ISSS), Berlin, Germany, August 4, 2015, see also http://systemology.org/manifesto.html

Rousseau, D., Wilby, J. M., Billingham, J., & Blachfellner, S. (2016a). A typology for the systems field. *Systema, Special Issue – General Systems Transdisciplinarity, 4*(1), 15–47.

Rousseau, D., Wilby, J. M., Billingham, J., & Blachfellner, S. (2016b). Manifesto for general systems transdisciplinarity. *Systema, Special Issue – General Systems Transdisciplinarity, 4*(1), 4–14.

Rousseau, D., Wilby, J. M., Billingham, J., & Blachfellner, S. (2016c). The scope and range of general systems transdisciplinarity. *Systema, Special Issue – General Systems Transdisciplinarity, 4*(1), 48–60.

Schindel, W. D. (2013). *Abbreviated Systematica™ 4.0 glossary—Ordered by concept*. Terre Haute, IN: ICTT System Sciences.

Schoderbek, P. P., Schoderbek, C. G., & Kefalas, A. G. (1990). *Management systems: Conceptual considerations* (Rev. ed.). Boston: IRWIN.

Sillitto, H., Dori, D., Griego, R. M., Jackson, S., Krob, D., Godfrey, P., et al. (2017). Defining "System": A comprehensive approach. *INCOSE International Symposium, 27*(1), 170–186.

Simon, H. A. (1962). The architecture of complexity. *Proceedings of the American Philosophical Society, 106*(6), 467–482.

Sutherland, J. W. (1973). *A general systems philosophy for the social and behavioral sciences*. New York: George Braziller.

Troncale, L. R. (1978). Linkage propositions between fifty principal systems concepts. In G. J. Klir (Ed.), *Applied general systems research* (pp. 29–52). New York: Plenum Press.

Van Fraassen, B. C. (2008). *Scientific representation: Paradoxes of perspective.* Oxford, UK: Oxford University Press.

von Bertalanffy, L. (1934). Wandlungen des biologischen Denkens. *Neue Jahrbücher Für Wissenschaft Und Jugendbildung, 10,* 339–366.

von Bertalanffy, L. (1950). The theory of open systems in physics and biology. *Science, 111*(2872), 23–29.

von Bertalanffy, L. (1955). An essay on the relativity of categories. *Philosophy of Science, 22*(4), 243–263.

von Bertalanffy, L. (1956). General system theory. *General Systems* [Article reprinted in Midgley, G. (Ed) (2003) *'Systems thinking'* (Vol. 1, pp 36–51). London: Sage. Page number references in the text refer to the reprint.], *1,* 1–10.

von Bertalanffy, L. (1969). *General system theory: Foundations, development, applications.* New York: Braziller.

von Bertalanffy, L. (1975). *Perspectives on general system theory.* New York: Braziller.

Wilby, J. M., Rousseau, D., Midgley, G., Drack, M., Billingham, J., & Zimmermann, R. (2015). Philosophical foundations for the modern systems movement. In M. Edson, G. Metcalf, G. Chroust, N. Nguyen, & S. Blachfellner (Eds.), Systems thinking: New directions in theory, practice and application, *Proceedings of the 17th Conversation of the International Federation for Systems Research, St. Magdalena, Linz, Austria, 27 April – 2 May 2014* (pp. 32–42). Linz, Austria: SEA-Publications, Johannes Kepler University.

Wimsatt, W. (2007). *Re-engineering philosophy for limited beings: Piecewise approximations to reality.* Cambridge, MA: Harvard University Press.

Chapter 3
The Potential of General Systemology as a Transdiscipline

Abstract The pioneers of the general systems movement envisioned the development of a new scientific 'meta-discipline' grounded in a "General Systems Theory" (GST*), a theory that encompasses universal principles underlying the systemic behaviours of all kinds of "real-world" systems. In contemporary terms we can identify this as a vision for a "transdiscipline" and we discuss its relationship to other conceptions of transdisciplinarity. In line with arguments presented elsewhere we identify this transdiscipline as "General Systemology", and the application of it as "General Systems Transdisciplinarity" (GSTD). The founders of the general systems movement argued that GSTD would be important for assisting the transfer knowledge between disciplines, facilitating interdisciplinary communication, supporting the development of exact models in areas where they are lacking, and promoting the "unity" of knowledge. In this chapter we defend this view, and infer that the scope and range of GSTD is wider than hitherto recognized, and argue that GSTD would potentially be the most powerful of the transdisciplines.

Keywords General Systemology · General Systems Transdisciplinarity · GSTD · Exploratory science · General system theory · GST · GST*

3.1 Introduction

The founders of the general systems movement (Ludwig von Bertalanffy, Kenneth Boulding, Anatol Rapoport and Ralph Gerard) called for the development of a new discipline with "interdisciplinary" and "meta-scientific" aspects (von Bertalanffy, 1972, p. 421), grounded in a "General Systems Theory (GST)" that would encompass the principles underlying the systemic behaviours of all kinds of systems (Boulding, 1956; Gerard, 1964; Rapoport, 1986; von Bertalanffy, 1950, 1969).

The term "General Systems Theory" has however come to stand for a wide range of meanings, including the foundational general theory of systems and the new meta-scientific discipline(s) which it would enable. In order to disambiguate between these meanings it has been proposed (Rousseau, Wilby, Billingham, & Blachfellner, 2016a), as discussed in Chap. 2, that:

© David Rousseau 2018

D. Rousseau et al., *General Systemology*, Translational Systems Sciences 13,
https://doi.org/10.1007/978-981-10-0892-4_3

- the foundational general theory of systems be called GST* (pronounced "g-s-t-star");
- the discipline that seeks to develop, apply and promote GST* be called "General Systemology" (in line with an earlier proposal by Bertalanffy scholars David Pouvreau and Manfred Drack (2007)); and
- the disciplinary field of systems, of which General Systemology is a component, be called "Systemology" (in line with an earlier proposal by Russ Ackoff (1973, p. 669).

We will employ this terminology in the present chapter.

The "Bertalanffy Circle" envisaged that this new "meta-science", would support interdisciplinary communication and cooperation, facilitate scientific discoveries in disciplines that lack exact theories, promote the unity of knowledge, and help to bridge the divide between the naturalistic and the human sciences (Laszlo, 1974, pp. 15–16, 19; Rapoport, 1976; von Bertalanffy, 1972, pp. 413, 423–424). The pioneers of General Systemology saw this as a strategy and action plan for averting looming social and environmental crises, and opening up a pathway towards a sustainable and humane future (Hofkirchner, 2005, p. 1; Laszlo, 1972; Pouvreau, 2014, p. 180).

General Systemology is still a nascent discipline (Francois, 2006, 2007), but interest in its development remains active (Billingham, 2014a; Friendshuh & Troncale, 2012; Rousseau & Wilby, 2014; Troncale, 2009). Recent times have seen an upsurge of interest in both GST* and General Systemology (Billingham, 2014a, 2014b; Denizan & Rousseau, 2014; Drack & Schwarz, 2010; Drack & Pouvreau, 2015; Hofkirchner & Schafranek, 2011; Rousseau, 2015b; Rousseau, Wilby, Billingham, & Blachfellner, 2015a) and we have recently called for concerted action towards advancing the development of this meta-science (Rousseau, Wilby, Billingham, & Blachfellner, 2015b, 2016b).

What von Bertalanffy called a "meta-science" would in current terms be called a "transdiscipline", and in fact Bertalanffy's call for a GST in the 1930s has attained a place in the history of transdisciplinarity as the moment when it became clear, as a scientific position, that objects and their environment interdepend, and hence that single disciplinary approaches would be insufficient for scientific analysis of complex problems (Balsiger, 2004, pp. 410–411). When the term "transdisciplinarity" was coined at the first conference on interdisciplinary research in 1970, one of the contributors, Jean Piaget, suggested that transdisciplinarity would support maturation and convergence between fields in terms of the general structures and fundamental patterns of thought, and that this would in turn lead to a general theory of systems (Apostel, Berger, & Michaud, 1972, p. 26). In his *Foreword* to a postmortem compilation of previously unpublished work by von Bertalanffy (edited by Edgar Taschdjian), Ervin Laszlo called General Systemology "a new paradigm for transdisciplinary synthesis" (von Bertalanffy, 1975, p. 12). We support these views, and in 2015 launched a *Manifesto* calling for the development of "General Systems

Transdisciplinarity (GSTD)" as representing the attitude and forms of action of General Systemology.

However, due to the rapid expansion of interest in transdisciplinarity the term "transdisciplinarity" has acquired a diversity of nuanced meanings, and so the term should be used with care.

The purpose of the present chapter is to explain some of the differences and commonalities between varieties of transdisciplinarity, clarify our perspective on the nature and value of transdisciplinarity, and present our vision for the potential reach and value of General Systemology understood as a transdiscipline.

3.2   What Is Transdisciplinarity?

The term "transdisciplinarity" was coined in a typology of terms devised at the first international conference on interdisciplinary research and teaching in OECD-member countries, held in Paris in 1970 (Apostel et al., 1972), where it was defined generically as "a common set of axioms for a set of disciplines". Since then interest in transdisciplinarity has grown rapidly, and it is currently "marked by an exponential growth of publications, a widening array of contexts, and increased interest across academic, public and private sectors" (Klein, 2014, p. 69).

3.2.1   The Scope of Transdisciplinarity

As a relatively new academic development there is as yet "no universal theory, methodology, or definition of transdisciplinarity" (Klein, 2013, p. 189)", and there is a considerable diversity of opinions about its nature, scope, value and potential. Sue McGregor called it a philosophical movement, (McGregor, 2014, p. 1) while Basarab Nicolescu identified it as a new kind of methodology (Nicolescu, 2002), but claimed it is not a new kind of discipline (Nicolescu, 2010a). Michael Gibbons and colleagues deny that it involves a methodology, but do claim that it is a new means of producing knowledge (Gibbons et al., 1994). According to both Predrag Cicovacki and McGregor, it requires a distinct axiological underpinning (Cicovacki, 2009; McGregor, 2014), but according to Nicolescu it does not (Nicolescu, 2005). Nicolescu has identified three kinds of transdisciplinarity which he classifies as respectively "theoretical transdisciplinarity" (which is concerned with developing transdisciplinary methodologies), "phenomenological transdisciplinarity" (which is concerned with using trans-disciplinary principles to build models and making predictions), and "experimental trans-disciplinarity" (which is concerned with doing experiments using transdisciplinary methodologies) (Nicolescu, 2010b, p. 23).

3.2.2   The Aims of Transdisciplinarity

Despite this diversity of views about the *nature* of transdisciplinarity, there is considerable coherence in claims about its *aims*. Klein indicated that it is about addressing unsolved problems, especially societal ones (Klein, 1986), Gibbons and colleagues say it is about joint efforts to address problems pertaining to the interplay between science, society and technology; problems that are not circumscribed in any existing disciplinary field (Gibbons et al., 1994). McGregor says it is an approach to solving deeply complex, interconnected problems that are too complex to be solved from within the boundaries of one discipline or by using a conventional empirical methodology (McGregor, 2014, pp. 3–4). For Seppo Tella, transdisciplinarity is intended to address the complex, wicked problems facing humanity (such as climate change, unsustainability, poverty) (Tella, 2005), and for McGregor it is about interconnecting science, politics and technology with society in a way that respects the survival of humanity in a future that is worth living (McGregor, 2014, pp. 2–3).

3.2.3   The Character of Transdisciplinarity

Klein's analysis of the discourses of transdisciplinarity (Klein, 2014) shows that all forms of transdisciplinarity engage with at least one of three overlapping concepts: transcendence, problem-solving, and transgression:

- "Transcendence" is about overcoming the barriers between disciplines, and in this sense transdisciplinarity is close to the ancient quest for the unity of knowledge, although the notion of "unity" has changed over time, to include aspects such as compatibility and consilience;
- Transdisciplinary approaches to problem-solving deviate from traditional approaches by placing great emphasis on "real world" problems, by involving feedbacks between organizations involved in research, design, education, services, and policymaking, and by a commitment to social, environmental, economic and ethically sustainable development; and
- "Transgression" is about questioning the constraints of traditional disciplines. This is not a rejection of the ethics or rationality of disciplinary inquiry, but an acknowledgement of uncertainty and a willingness to critique, reimagine, reframe or reformulate the status quo. This attitude allows established boundaries and limitations to be challenged and existing knowledge to be recontextualised, and in so doing opens up new routes to discovery, insight, and innovation.

3.2.4 The Varieties of Transdisciplinarity

Transdisciplinarity is currently a dappled arena, with much consistency in its overall aims but also much diversity in how those aims are pursued. Klein has explained that "transdisciplinarity is simultaneously an attitude and a form of action" (Klein, 2004, p. 521). This characterisation is helpful in understanding the diversity of forms transdisciplinarity currently takes, when taken together with the definition of transdisciplinarity as "a common set of axioms for a set of disciplines". There are many kinds of "axioms" that can be proposed as assumptions, beliefs or principles that would, if adopted, lead to the kind of "better world" that transdisciplinarity is focused on. The "attitudes" that inform transdisciplinary "forms of action" have ranged over distinct frameworks such as general systems, post-structuralism, Marxism, feminist theory, cultural critique and sustainability (Klein, 2014).

This diversity highlights a key question for transdisciplinarity, namely whether it represents a discipline in its own right or merely modulates the attitude with which existing disciplinary work is undertaken. We propose that this issue could be resolved in the light of the systemic model of an academic discipline we have presented elsewhere (Rousseau, Wilby, et al., 2016a) and discussed in Chap. 2. This represents a discipline as an "Activity Scope" informed by a "Knowledge Base" and conditioned by a "Guidance Framework", which we call "the AKG model" for short.

The AKG model provides a way of distinguishing between a topic, a theory, an activity, an attitude and a complete discipline. In the light of this model we can see that the current diversity of kinds of transdisciplinarity can be characterised in terms of two major types. The first type involves a concern for the application of specific transdisciplinary values such as equal opportunity or sustainability. These kinds of values can be applied across multiple disciplines, but this serves only to extend the guidance frameworks of existing disciplines rather than generating transdisciplines as such. In the second type, transdisciplinarity involves the application, under a guidance framework (which includes values), of transdisciplinary theories such as GST* or Cybernetics. For this second type it is appropriate to speak of transdisciplinarity as the application of a transdiscipline, since there is a distinct discipline involved *in addition to* the orthodox ones over which its applicability might range.

In this light we can not only understand the origins of the diversity of kinds of transdisciplinarity that we have today, but we can see that the first type of disciplinarity is likely to evolve into the second type, as its proponents firstly develop methodologies for applying those value systems in different disciplinary contexts, and as theories are developed that explain the utility or appropriateness of those values and hence ground those methodologies in principled ways. From this we can view "type 1" transdisciplinarity as "early-stage type 2" transdisciplinarity, and see its evolution from "type 1" to "type 2" as a maturation from an intuitively compelling form of activism to an objectively compelling species of scientific endeavour.

However, we can also see that the value systems of current "type 2" transdisciplines will increasingly evolve under the influence of "type 1" transdisciplinarity to include transdisciplinary values, shifting them further from the classical ideal of science as a "value-neutral" endeavour to one that accepts responsibility for its impact in the world. We can thus foresee an evolutionary trajectory for all kinds of transdisciplinarity, involving the development of transdisciplines that incorporate transdisciplinary theories, methodologies and values. Transdisciplinarity can begin with any of these three components, but in viable forms of transdisciplinarity all three components will eventually be present. Moreover, we can anticipate that on the basis of an emerging consilience between transdisciplinary theories, methodologies and values the diverse transdisciplines might coalesce into a coherent transdisciplinary field. For the purposes of this chapter we will henceforth discuss transdisciplinarity only in terms of an "ideal type" that (potentially if not yet actually) is the expression of a transdiscipline involving transdisciplinary theories, methodologies and values, and whose values align with a concern for building a "better world".

3.3 Kinds of Disciplinarity

The focus of transdisciplinarity on problem solving calls for an explanation of how transdisciplinarity differs from other kinds of disciplinarity in its approach to problem solving, and how its particular value arises. Several kinds of disciplinarity are now recognised (see, e.g. Klein, 2010; Nicolescu, 2010a; Salmons & Wilson, 2007; Tress, Tress, & Fry, 2005), as follows:

1. *Mono-disciplinarity*: this involves only a single discipline and is suitable for addressing well-bounded phenomena or a single aspect of a complex phenomenon;
2. *Multi-disciplinarity*: this is used for addressing multiple aspects of a phenomenon by making use of several disciplines. It acknowledges their differences but involves no attempt to bridge between them;
3. *Cross-disciplinarity*: this is used where several academic disciplines are interested in the same aspect of a complex phenomenon. The different disciplines' distinct methods are brought to bear on the same problem in a coordinated way, establishing a kind of middle ground;
4. *Inter-disciplinarity*: this involves combining several disciplines, attempting to synthesize them into something that provides a new perspective on the given problem; and

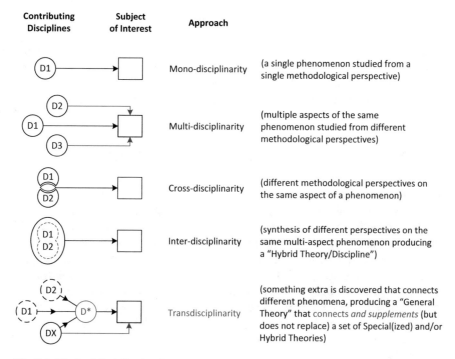

Fig. 3.1 Kinds of disciplinarity. (Reproduced from (Rousseau, Wilby, Billingham, & Blachfellner, 2016c), with permission)

5. *Transdisciplinarity*: this involves disciplinary frameworks that are developed from generalisations based on patterns[1] that recur across or connect between several disciplines, and hence it involves insights about the general nature of the world rather than the special natures of specific kinds of phenomena (Klein, 2004, p. 515; Rousseau & Wilby, 2014). In contrast to other kinds of disciplinarity which bring the means of one or more specialised disciplines to bear on a specific problem, transdisciplinary frameworks are relevant to the phenomena studied in several disciplines, and hence transdisciplinarity introduces new means that can enhance the effectiveness of the disciplines it is partnered with (Nicolescu, 2002, pp. 44, 46).

These distinctions are illustrated in Fig. 3.1. Here the specialised disciplines are indicated by numbers D1-D3, the meta-discipline representing trans-disciplinarity by D*, and any form of disciplinarity *other than* transdisciplinarity by DX.

The kinds of disciplinarity illustrated in Fig. 3.1 follow a progression of increasing complexity and power beyond that of the specialized mono-disciplines,

---

[1] The term "patterns" here can be taken to range over both concrete patterns such as patterns of structures, processes and behaviours in natural systems, or abstract patterns observed in social systems, value systems and patterns of thought.

reflecting engagement with increasingly complex problems or increasingly deep questions.

Note that transdisciplinarity is different from the others in that it adds something new to the disciplines it generalises over, rather than combining or merging existing disciplinary resources. Its value is realised when it is used in conjunction with one of those disciplines to address problems originating in those disciplines.

3.4   The Range of General Systems Transdisciplinarity

In general, different kinds of disciplinarity are called for depending on the complexity of the phenomena being addressed. The problems any discipline tries to address can be divided into three broad categories, as shown in Fig. 3.2, namely "Routine" (problems that can readily addressed by established explanatory models and theories), "Difficult" (problems involving phenomena not yet understood but that appear analysable using existing research methods) and "Radical" (problems involving phenomena that cannot be analysed given the kinds of theories/methods available in the discipline). These are provisional categories, because it is never certain that a given phenomenon has been correctly or fully explained, is really readily analysable, or actually does lie outside the analytic capacity of the given disciplinary framework.

In every discipline the central objective is to maximize the scope of what can be explained, predicted, managed or utilized. Doing this calls for different kinds of disciplinarity depending on the complexity of the issue, as shown in Fig. 3.2. When dealing with a specific challenge the kinds of disciplinarity are typically engaged in the order of their relative complexity, in order to find the solution in the simplest possible way. However, given the nature and range of phenomena that still lie beyond scientific explanation, it is likely that scientific investigation will increasingly call for transdisciplinary working.

Transdisciplinarity is grounded in insights about patterns that recur across or connect between disciplines, and therefore it tells us something about the fundamental nature of the world that is not readily evident from within the specialized disciplines. Because of this it can powerfully enhance problem solving techniques in specialised areas, and thus be especially useful where specialised disciplines are addressing apparently intractable disciplinary problems, such as those that reflect deep ontological or epistemic issues.

Amongst the transdisciplines, General Systemology is arguably the potentially most powerful, because it is grounded in the deepest of the general principles applying to the "real" world, as we have discussed elsewhere (Rousseau, 2015a; Rousseau, Billingham, Wilby, & Blachfellner, 2016d; Rousseau, Wilby, et al., 2016a) and in Chaps. 2 and 4. We will present some example principles in Chap. 6, where, for example, we present a general scientific systems principle analogous to the principle of conservation of energy as used in science more generally. Just like conservation of energy the principles of General Systemology will represent insights that are

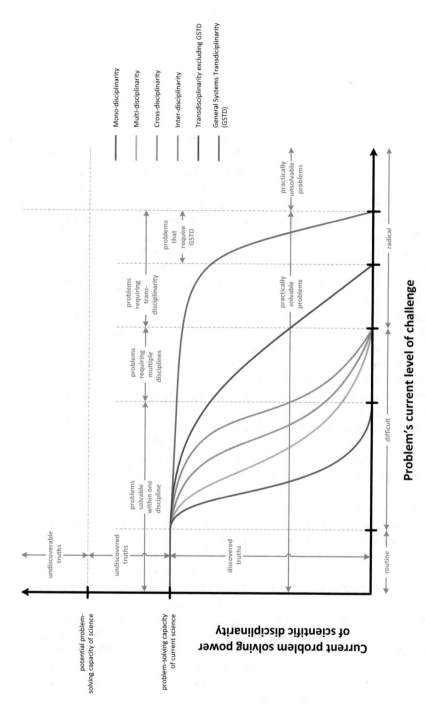

Fig. 3.2 The application areas of kinds of disciplinarity. (Reproduced from (Rousseau, Wilby, et al., 2016c), with permission)

relevant in all disciplines and in all contexts. However, some of them will have application beyond the principles of science, applying also, for example to abstract and conceptual systems.

3.5   The Scope of General Systems Transdisciplinarity

Figure 3.2 illustrates our claim that GSTD is more versatile than other forms of transdisciplinarity. This is so because General Systemology seeks to identify universal principles underlying the origin, evolution and behaviour of all kinds of complex systems (Billingham, 2014a; Friendshuh & Troncale, 2012; Rousseau, 2014a; Rousseau et al., 2015a). As such its concepts, models and methodologies could be relevant in all areas of investigation and theory development. The transdisciplinary insights of General Systemology might be used not only to address complex problems, but also to support exploratory science, i.e. to develop testable hypotheses about unexplained complex phenomena that are not considered to be problematic but are nevertheless part of the context in which problem-solving is undertaken. For example, many familiar human abilities such as creativity and abstract thinking remain largely mysterious, and yet understanding them would contribute much to achieving the thrivable future that is the focus of transdisciplinary ambitions.

The transdisciplinary nature of General Systemology also positions it uniquely to support translational science. From a very wide perspective science can be seen as supporting a cycle which addresses human needs by seeking discoveries that can be turned into insights that can support innovation towards tools and practices that fulfil those needs. Translational science is concerned with expediting the transition steps from discovery to insight to innovation to application, and from need identification to exploration for discovery (Drolet & Lorenzi, 2011; Woolf, 2008). We illustrate the scope of the translational science challenge in Fig. 3.3. The arrows indicate the interfaces between the different groups that contribute to the disciplinary activity cycle, and every one of these interfaces holds a translational challenge. The diagram shows that the scope of the translational problem is much wider than just the transition from engineering to practice, where the majority of current effort and funding is focused, and necessarily cuts across all the traditional disciplines.

Translational Science faces many challenges (Fang & Casadevall, 2010), and it has been suggested that the systems perspective can enhance the effectiveness of Translational Science (Kijima, 2015). The founders of General Systemology already envisaged that General Systemology would facilitate co-operation and transfers of knowledge between disciplines via its common framework of general concepts, principles and models. This would increase the effectiveness of disciplines by reducing duplication of effort. We can however now see that GSTD would also facilitate communication and cooperation between research organizations, academic institutions, industries and practitioners within a discipline. In this way GSTD can also help to accelerate the cycle whereby discoveries lead to innovations and those in turn become products and services. In sense GSTD would provide

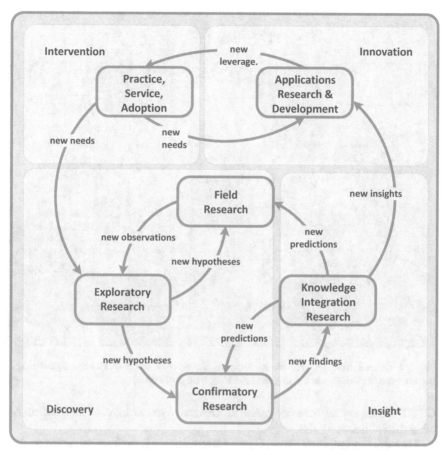

Fig. 3.3 Disciplinary activity groups and their interfaces. (Adapted from (Rousseau, Wilby, et al., 2016a))

essential support for the development of the currently emerging Translational *Systems* Science.

The way in which GSTD can support these new developments is illustrated in Fig. 3.4. We have here used the same colour scheme as we did for the AKG Model of a discipline we showed in Chap. 2, using blue for components of the Knowledge Base, orange for components of the Guidance Framework, and green for components of the Activity Scope. The diagram illustrates the key components of General Systemology and shows the scope of its activities. As can be seen in the diagram, the activity scope of General Systemology has two transdisciplinary aspects. In the first, shown in the left half of the diagram, General Systemology functions as the unifying transdiscipline for Systemology, refining and extending the general theory (GST*) that applies across the specialized and hybrid systems disciplines, as we discuss in Chap. 2. In the second aspect, shown in the right half of the diagram,

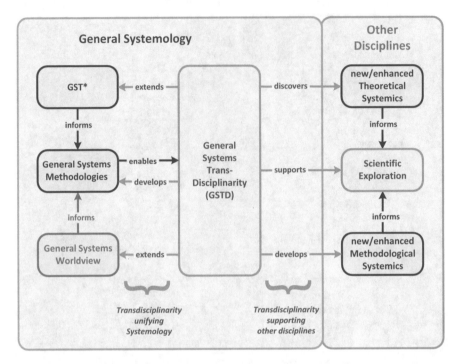

Fig. 3.4 General Systems Transdisciplinarity as the activity scope of General Systemology. (Reproduced from (Rousseau, Wilby, et al., 2016c), with permission)

GSTD leverages the methodologies of General Systemology to support/extend other disciplines and fields.

Amongst the transdisciplines, General Systemology is perhaps the only one that has a *scientific strategy* for finding transdisciplinary patterns, by following von Bertalanffy's injunction to look for isomorphies of structures, behaviors and processes present in the designs of different kinds of systems under the guidance of the GSW, as we discuss in Chap. 4. As a worldview GSW embraces the framework of Broad Naturalism and Moderate Critical Realism that is dominant in the sciences. However, it must be noted that unlike the science ideal of neutrality, General Systemology has from the outset maintained a concern for meaning and value (Hammond, 2003, 2005) and a commitment to building a "better world" (Hofkirchner & Schafranek, 2011). As such it has always pursued the ambition of bridging the gap between the object-oriented and the subject-oriented disciplines in a way that preserves the merits of each, and recent developments in General Systemology suggest that such a bridge can in fact be attained via the development of GST* and the GSW (Rousseau, 2014b, 2015a). In this light, General Systemology is likely to contribute significantly to the discovery, problem-solving and cultural transformation that will be needed to help us attain and sustain a thriving eco-civilization.

3.6 Summary

In this chapter we explored the differences between kinds of disciplinarity, including mon-, multi- cross-, inter- and transdisciplinarity, and reflected on the value of each. We pointed that at present there are multiple kinds of transdisciplinarity, but argued that these reflect differences in evolutionary trajectories and they can be expected to converge (or at least become consilient) as transdisciplinary theories become more mature, and as links between them become evident on the basis of advances in GST*. In this way, we foresee the development of a general systems transdisciplinarity (GSTD) that will have relevance in all areas of human and scientific inquiry, and provide a means to explore and address deep problems beyond the current scope of other kinds of disciplinarity. Given the potential value of GSTD, we have worked on, and presented elsewhere (Rousseau, Blachfellner, Wilby & Billingham, 2016), a detailed Research Agenda outlining the issues to be addressed towards achieving such a profound and empirically significant general systems transdisciplinarity. Some of these ideas for key research focus have already received some attention, and that progress is presented in the Chapters following.

References

Ackoff, R. L. (1973). Science in the systems age: Beyond IE, OR, and MS. *Operations Research, 21*(3), 661–671.

Apostel, L., Berger, G., & Michaud, G. (1972). *Interdisciplinarity: Problems of teaching and research in universities*. Paris: OECD.

Balsiger, P. W. (2004). Supradisciplinary research practices: History, objectives and rationale. *Futures, 36*(4), 407–421.

Billingham, J. (2014a). *GST as a route to new systemics*. Presented at the 22nd European Meeting on Cybernetics and Systems Research (EMCSR 2014), Vienna, Austria. In J. M. Wilby, S. Blachfellner, & W. Hofkirchner (Eds.), EMCSR 2014: Civilisation at the crossroads – Response and responsibility of the systems sciences, Book of Abstracts (pp. 435–442). Vienna: EMCSR.

Billingham, J. (2014b). *In search of GST*. Position paper for the 17th conversation of the International Federation for Systems Research on the subject of 'philosophical foundations for the modern systems movement', St. Magdalena, Linz, Austria, 27 April–2 May 2014, pp. 1–4.

Boulding, K. E. (1956). General systems theory — The skeleton of science. *Management Science, 2*(3), 197–208.

Cicovacki, P. (2009). Transdisciplinarity as an interactive method: A critical reflection on the three pillars of transdisciplinarity. *Integral Leadership Review*, 9(5). Retrieved from http://integral-leadershipreview.com/4549-feature-article-transdisciplinarity-as-an-interactive-method-a-critical-reflection-on-the-three-pillars-of-transdisciplinarity/

Denizan, Y., & Rousseau, D. (2014). Bertalanffy and beyond: Improving Systemics for a better future. A symposium of the EMCSR 2014, 22-25 April. In *Civilisation at the crossroads: Response and responsibility of the systems sciences* (pp. 409–410). Vienna: BCSSS. Retrieved from http://emcsr.net/calls-2014/calls-for-papers-2014/bertalanffy-and-beyond-improving-systemics-for-a-better-future/

Drack, M., & Pouvreau, D. (2015). On the history of Ludwig von Bertalanffy's "general Systemology", and on its relationship to cybernetics – Part III: Convergences and divergences. *International Journal of General Systems, 44*(5), 523–5571.

Drack, M., & Schwarz, G. (2010). Recent developments in general system theory. *Systems Research and Behavioral Science, 27*(6), 601–610.

Drolet, B. C., & Lorenzi, N. M. (2011). Translational research: Understanding the continuum from bench to bedside. *Translational Research, 157*(1), 1–5.

Fang, F. C., & Casadevall, A. (2010). Lost in translation – Basic science in the era of translational research. *Infection and Immunity, 78*(2), 563–566.

Francois, C. (2006). Transdisciplinary unified theory. *Systems Research and Behavioral Science, 23*(5), 617–624.

Francois, C. (2007). *Who knows what general systems theory is?* Retrieved January 31, 2014, from http://isss.org/projects/who_knows_what_general_systems_theory_is

Friendshuh, L., & Troncale, L. R. (2012, July 15–20). *Identifying fundamental systems processes for a general theory of systems.* In Proceedings of the 56th annual conference, International Society for the Systems Sciences (ISSS), San Jose State University, 23 pp.

Gerard, R. W. (1964). Entitation, Animorgs and other systems. In M. D. Mesarović (Ed.), *Views on General System Theory: Proceedings of the 2nd systems symposium at Case Institute.* New York: Wiley.

Gibbons, M., Limoges, C., Nowotny, H., Schwartzman, S., Scott, P., & Trow, M. (1994). *The new production of knowledge: The dynamics of science and research in contemporary societies.* Los Angeles: Sage.

Hammond, D. (2003). *The science of synthesis: Exploring the social implications of general systems theory.* Boulder, CO: University Press of Colorado.

Hammond, D. (2005). Philosophical and ethical foundations of systems thinking. *TripleC: Communication, Capitalism & Critique, 3*(2), 20–27.

Hofkirchner, W. (2005). Ludwig von Bertalanffy, Forerunner of evolutionary systems theory. In *The new role of systems sciences for a knowledge-based society.* Proceedings of the First World Congress of the International Federation for Systems Research, Kobe, Japan, CD-ROM (Vol. 6, ISBN 4-903092-02-X).

Hofkirchner, W., & Schafranek, M. (2011). General System Theory. In C. A. Hooker (Ed.), *Vol. 10: Philosophy of complex systems* (1st ed., pp. 177–194). Amsterdam: Elsevier BV.

Kijima, K. (2015). Translational and trans-disciplinary approach to service systems. In *Service systems science* (pp. 37–54). Springer.

Klein, J. T. (1986). Postscript: The broad scope of interdisciplinarity. In D. E. Chubin, A. L. Porter, F. A. Rossini, & C. Terry (Eds.), *Interdisciplinary analysis and research: Theory and practice of problem-focused research and development* (pp. 409–424). Mount Airy, MD: Lomond Publications.

Klein, J. T. (2004). Prospects for transdisciplinarity. *Futures, 36*(4), 515–526.

Klein, J. T. (2010). The taxonomy of interdisciplinarity. In R. Frodeman, J. T. Klein, & C. Mitcham (Eds.), *The Oxford handbook of interdisciplinarity* (pp. 15–30). Oxford: Oxford University Press. Retrieved from http://philpapers.org/rec/KLETOH

Klein, J. T. (2013). The transdisciplinary moment (um). *Integral Review, 9*(2), 189–199.

Klein, J. T. (2014). Discourses of transdisciplinarity: Looking back to the future. *Futures, 63*, 68–74.

Laszlo, E. (Ed.). (1972). *The relevance of general systems theory.* New York: George Braziller.

Laszlo, E. (1974). *A strategy for the future.* New York: Braziller.

McGregor, S. L. (2014, April–June 2014). Transdisciplinary axiology: To be or not to be? *Integral Leadership Review.*

Nicolescu, B. (2002). *Manifesto of transdisciplinarity* (K.-C. Voss, Trans.). Albany: SUNY Press.

Nicolescu, B. (2005). *Transdisciplinarity – Past, present and future.* In II Congresso Mundial de Transdisciplinaridade. Vila Velha/Vitória, SC, Brasil: CETRANS. Retrieved from http://cetrans.com.br/textos/transdisciplinarity-past-present-and-future.pdf

Nicolescu, B. (2010a). *Disciplinary boundaries – What are they and how they can be transgressed?* In International symposium on Research across boundaries. Luxembourg: University of Luxembourg.

Nicolescu, B. (2010b). Methodology of transdisciplinarity–levels of reality, logic of the included middle and complexity. *Transdisciplinary Journal of Engineering & Science, 1*(1), 19–38.

Pouvreau, D. (2014). On the history of Ludwig von Bertalanffy's "general systemology", and on its relationship to cybernetics – Part II: Contexts and developments of the systemological hermeneutics instigated by von Bertalanffy. *International Journal of General Systems, 43*(2), 172–245.

Pouvreau, D., & Drack, M. (2007). On the history of Ludwig von Bertalanffy's "general Systemology", and on its relationship to cybernetics, part 1. *International Journal of General Systems, 36*(3), 281–337.

Rapoport, A. (1976). General systems theory: A bridge between two cultures. Third annual Ludwig von Bertalanffy memorial lecture. *Behavioral Science, 21*(4), 228–239.

Rapoport, A. (1986). *General System Theory: Essential concepts & applications.* Cambridge, MA: Abacus.

Rousseau, D. (2014a). *Foundations and a framework for future waves of systemic inquiry.* Presented at the 22nd European meeting on cybernetics and systems research (EMCSR 2014), Vienna, Austria. In J. M. Wilby, S. Blachfellner, & W. Hofkirchner (Eds.), EMCSR 2014: Civilisation at the crossroads – response and responsibility of the systems sciences, book of abstracts (pp. 428–434). Vienna: EMCSR.

Rousseau, D. (2014b). Reconciling spirituality with the naturalistic sciences: A systems-philosophical perspective. *Journal for the Study of Spirituality, 4*(2), 174–189.

Rousseau, D. (2015a). General systems theory: Its present and potential [Ludwig von Bertalanffy memorial lecture 2014]. *Systems Research and Behavioral Science, Special Issue: ISSS Yearbook, 32*(5), 522–533.

Rousseau, D. (2015b, May). *Systems philosophy and its relevance to systems engineering.* Webinar presented as Webinar 76 of the International Council on Systems Engineering.

Rousseau, D., Billingham, J., Wilby, J. M., & Blachfellner, S. (2016). The synergy between general systems theory and the general systems worldview. *Systema, Special Issue – General Systems Transdisciplinarity, 4*(1), 61–75.

Rousseau, D., Blachfellner, S., Wilby, J. M., & Billingham, J. (2016). A research agenda for general systems Transdisciplinarity (GSTD). *Systema, Special Issue - General Systems Transdisciplinarity, 4*(1), 93–103.

Rousseau, D., & Wilby, J. M. (2014). Moving from disciplinarity to transdisciplinarity in the service of thrivable systems. *Systems Research and Behavioral Science, 31*(5), 666–677.

Rousseau, D., Wilby, J. M., Billingham, J., & Blachfellner, S. (2015a, January 24–27, 2015). *In search of General Systems Transdisciplinarity.* Presented at the International Workshop of the Systems Science Working Group (SysSciWG) of the International Council on Systems Engineering (INCOSE), Torrance, Los Angeles.

Rousseau, D., Wilby, J. M., Billingham, J., & Blachfellner, S. (2015b, August 4, 2015). *Manifesto for General Systems Transdisciplinarity (GSTD).* Plenary Presentation at the 59th Conference of the International Society for the Systems Sciences (ISSS), Berlin, Germany, see also http://systemology.org/manifesto.html

Rousseau, D., Wilby, J. M., Billingham, J., & Blachfellner, S. (2016a). A typology for the systems field. *Systema, Special Issue – General Systems Transdisciplinarity, 4*(1), 15–47.

Rousseau, D., Wilby, J. M., Billingham, J., & Blachfellner, S. (2016b). Manifesto for general systems transdisciplinarity. *Systema, Special Issue – General Systems Transdisciplinarity, 4*(1), 4–14.

Rousseau, D., Wilby, J. M., Billingham, J., & Blachfellner, S. (2016c). The scope and range of general systems transdisciplinarity. *Systema, Special Issue – General Systems Transdisciplinarity, 4*(1), 48–60.

Salmons, J., & Wilson, L. (2007). *Crossing a line: An interdisciplinary conversation about working across disciplines.* A trainerspod.com webinar: August 23, 2007.

Tella, S. (2005). Multi-, inter-and transdisciplinary affordances in foreign language education: From singularity to multiplicity. In J. Smeds, K. Sarmavouri, E. Laakkonen, & R. De Cillia (Eds.), *Multicultural communities, multicultural practice* (pp. 67–88). Turku: University of Turku.

Tress, G., Tress, B., & Fry, G. (2005). Clarifying integrative research concepts in landscape ecology. *Landscape Ecology, 20*(4), 479–493.

Troncale, L. R. (2009). Revisited: The future of general systems research: Update on obstacles, potentials, case studies. *Systems Research and Behavioral Science, 26*(5), 553–561.

von Bertalanffy, L. (1950). An outline of general system theory. *British Journal for the Philosophy of Science, 1*(2), 134–165.

von Bertalanffy, L. (1969). *General system theory: Foundations, development, applications.* New York: Braziller.

von Bertalanffy, L. (1972). The history and status of general systems theory. *Academy of Management Journal, 15*(4), 407–426.

von Bertalanffy, L. (1975). *Perspectives on general system theory.* New York: Braziller.

Woolf, S. H. (2008). The meaning of translational research and why it matters. *JAMA, 299*(2), 211–213.

Chapter 4
The Existence, Nature and Value
of General Systems Theory (GST*)

Abstract The founders of the general systems movement envisaged the development of a theory articulating and inter-relating the principles underlying the systemic behaviours of all kinds of concrete systems. We call this theory GST* ("g-s-t-star") to disambiguate it from other uses of the term "GST" prevalent in the literature. GST* is still radically underdeveloped, but its nature can be analysed. GST* is a formal theory, because the principles of GST* would apply across all kinds of systems, that is, GST* would predict behaviours and structures of systems qua systems, without regard for the kind of system under consideration, and hence it is neutral with respect to ontology.

There is a long-standing controversy within the systems community about whether a GST* exists in principle, whether it would be of practical value if it did, and how its principles might be discovered. In this chapter we argue by analogy from the history of science that if a GST* could be developed it would be highly valuable, and show that its existence is predicated on the assumption of a philosophical framework called the General Systems Worldview (GSW). We present an argument that development of the General Systems Worldview can guide us to the discovery of general systems principles for a GST*, and that together GST* and the GSW can ground the development of a powerful General Systems Transdiscipline, now called "General Systemology".

Keywords General systems theory · GST · GST* · Systems philosophy · General systems worldview · General Systemology

4.1 Introduction

The *Society for the Advancement of General Systems Theory* was founded in 1954, incorporated as the *Society for General Systems Research* (SGSR) in 1956, and lives on today as the *International Society for the Systems Sciences*(ISSS) (since 1988). A central ambition of the founders was to develop a theory encompassing the principles that apply to systems in general. However, progress has been slow, and even after 60 years of effort GST* remains underdeveloped, especially in its foundational aspects (Wilby et al., 2015; Francois, 2007; Pouvreau, 2013, p. 864).

D. Rousseau et al., *General Systemology*, Translational Systems Sciences 13, https://doi.org/10.1007/978-981-10-0892-4_4

From the outset there was a wide ranging set of views about whether such a theory existed in principle, and what it might entail, and this debate is still ongoing.

First, some doubt that such a theory could exist at all. For example, Mario Bunge has claimed that that there cannot be a GST*, but only theories about special kinds of systemic structures and behaviours ("Systemics") that are unified by a philosophical framework (Bunge, 1979, p. 1, 2014, p. 8).

Second, it has been claimed, for example by Boulding and Herbert Simon, that if a GST* existed it would not be of much practical significance, on the basis that "we always pay for generality by sacrificing content, and all we can say about practically everything is almost nothing" (Boulding, 1956a, p. 128), and "systems of these diverse kinds could hardly be expected to have any non-trivial properties in common" (Simon, 1996, p. 173).

Third, it was radically unclear how the process of discovering a GST might proceed. Von Bertalanffy and his 'circle' had few practical ideas about how to go about discovering this theory (Checkland, 1993, p. 93; Dubrovsky, 2004, p. 110; Pouvreau, 2014, p. 180), and as we noted earlier very limited progress has been made so far. Instead, what has happened is that the practical offshoots of theories addressing specific systemic behaviours or structures, originally foreseen as forebears of GST*, have gained prominence (Flood & Robinson, 1989, p. 63), while the general isomorphies they represent have not been assimilated into a general theory (Francois, 2004, p. 248).

The main objectives of this chapter are to answer these three doubts, by (a) showing a philosophical basis on which one can argue for the existence, in principle, of a GST*, (b) arguing that GST* would be of great practical significance if used in tandem with an appropriate philosophical framework, and (c) showing that Systems Philosophy can aid the development of GST* in a principled manner. On this basis we foresee the development of General Systemology as a practical and desirable prospect.

4.2   The Potential Utility of GST*

Assertions about the lack of utility GST* would have as a general theory are contradicted by examples from the history of science. Darwin's *Theory of Evolution by Natural Selection*, Mendeleev's *Periodic Table of the Chemical Elements*, Newton's *Laws of Mechanics* and Lyell's *Principles of Geology* transformed their respective disciplines by unifying hitherto fragmented areas of study under a common conceptual and explanatory framework, thereby rapidly opening up new avenues to scientific discovery.

Given the examples of history, it seems reasonable to suggest that if a GST* existed it is likely to be as important in relation to the study of kinds of systems as these other theories have been to their respective subject matters. Extrapolating from these analogies it is plausible (as we have argued elsewhere (Billingham,

2014a, 2014b; Rousseau, Billingham, Wilby, & Blachfellner, 2016a) and as we further discuss in Chap. 5), that a mature GST* will unify the systems field by providing both a 'gestalt' that relates the special theories describing the specific systemic behaviours and structures that occur in Nature to each other, and the principles that entail their evolution in Nature. It will be some time yet before this fully developed GST* is available, but in the meantime even a partial GST* is likely to be an important stimulus for discovering new kinds of systemic structures and behaviours, just as the partial Periodic Table was an important guide and inspiration for the discovery of new chemical elements.

Insofar as specific systemic structures and behaviours are modelled by the special theories collectively known as "Systemics" (Bunge, 1979, p. 1), the implication is that the development of GST* will provide a principled basis for the discovery of new Systemics via General Systemology, as opposed to the incidental way in which Systemics have been discovered to date within the specialised disciplines. This is an exciting prospect because it entails not only the discovery of new ways to understand, design, engineer or govern systems, but it means that General Systemology, informed by GST*, will reveal systemic structures and mechanisms unknown to and unanticipated by contemporary science. Progress towards a GST* will therefore not only unify the systems field but initiate an important cycle of scientific discovery. However, this positive outlook is subject to two significant caveats.

First, the examples given earlier of unifying theories offered as analogies are all empirical theories about some aspect of concrete nature, whereas GST* is a formal theory that generalizes over the special systems theories, themselves generalizations over multiple disciplines. The "meaning" of GST* in any specific empirical context is thus unclear until it is combined with an appropriate philosophical framework representing the concrete world in terms of the systems paradigm. The philosophical framework of the "Bertalanffy Circle" embraced what we would today characterize as Broad Naturalism and Moderate Scientific Realism. This framework, which von Bertalanffy called "Perspectivism",[1] underpins a worldview that has been called the "General Systems Worldview" (GSW) (Elohim, 2000; von Bertalanffy, 1934), and articulating it is one of the objectives of Systems Philosophy, the philosophical element of Systemology. However, it is acknowledged that GST* could equally well be interpreted in terms of other philosophical stances, and may have other kinds of utility in those cases.

Second, the extent of the value of GST* depends on a very strong philosophical claim, namely that every concrete thing is a system or part of one (Bunge, 1979, p. 44, 245). This is a core tenet of the GSW, and if this assumption is true then GST* would be relevant in all cases where science is studying concrete phenomena. In this case, having a GST* would be enormously empowering to all the specialized disciplines. Investigating the validity of the assumption that everything is a system or part of one must therefore be one of the core objectives of a research agenda for General Systemology.

---

[1] See Chap. 2 note 4 for more on Perspectivism.

4.3   The Potential Existence of GST*

4.3.1   Recent Developments in General Systemology

As mentioned earlier, we have only fragments of a GST* today, despite occasional claims to the contrary. However, several researchers have recently tried to open up new routes forward, for example Len Troncale is developing an approach based on (inter alia) linkage propositions between systems isomorphies (Friendshuh & Troncale, 2012), Kevin Adams and colleagues are developing an approach based on systems axioms (Adams, Hester, Bradley, Meyers, & Keating, 2014), and Ion Georgiou is working on alternative epistemological approaches (Georgiou, 2006).

Our own approach, and a central focus of this chapter, is based on Systems Philosophy, a transdiscipline formally founded by Ervin Laszlo (1972a) but representing a tradition going back to the pre-Socratic philosophers (Laszlo, 1972a, p. 291; Pouvreau, 2014, p. 206). The central focus of Systems Philosophy is to develop a worldview based on scientific principles and the systems paradigm, and to use it to solve important problems in science, philosophy and society (Laszlo, 1972a, p. 12). The connection we aim to pursue is that, as Laszlo pointed out, there is an intimate relationship between this worldview and GST*, which we discuss later on. We do not yet have a fully-fledged version of this worldview either, but the situation is much more advanced than is the case for GST*. The worldview at stake here is informed by the findings of science and philosophy of science as well as by the systems paradigm, and so has much material to draw on. This perspective is traditionally called the "General Systems Worldview" (GSW), and we will maintain that convention.

We will argue that the tenets of the GSW entail the existence of a GST*, that the development of the GSW can make important contributions to the development of GST*, and that progress with GST* will in turn inform the refinement of the GSW. To prepare the ground for presenting these arguments, a closer look at the notion of a "worldview" is needed.

4.3.2   Worldview as a Perspective on the World and on Life

The term "worldview" is the English rendering of the term *Weltanschauung*. It was coined by Immanuel Kant in 1790 (Kant & Gregor, 1987, pp. 111–112), and it rapidly developed as "a term for an intellectual conception of the universe from the perspective of a human knower" (Naugle, 2002, p. 59).

The term worldview has a rich academic history, and a dappled application of terminology, which we will not review here; for comprehensive surveys see (Hiebert, 2008; Naugle, 2002; Sire, 2004). Essentially, a worldview is a "map of reality" that people use to order their lives (Hiebert, 2008, p. 15).

A worldview can be characterized as comprising three main elements, namely a perspective on the nature of knowledge ("epistemology"), a perspective on the objective nature of the universe (a "world picture" or *Weltbild*) and a perspective on the subjective significance of one's existence in the world (a "life view" or *Lebensanschauung*) (Hofkirchner, 2013, p. 38). More technically and in more detail, we can define a worldview in contemporary terms as encompassing the following components:

1. *An Epistemology* (theory about what kinds of knowledge are possible and how to gain knowledge);
2. *An Ontology* (model of what exists most fundamentally);
3. *A Metaphysics* (model of the nature of what exists, i.e. what is possible given the Ontology);
4. *Cosmology* (high-level theory of the origins, history, organisation and destiny of the world);
5. *Axiology* (value system and theories about what is important and why); and
6. *Praxeology* (theory about the nature of action, agency, freedom and responsibility).

In this list, Ontology, Metaphysics and Cosmology comprise the objective "world picture" and Axiology and Praxeology comprise the subjective "life view".

4.3.3   The Foundational Tenets of the General Systems Worldview (GSW)

The General Systems Worldview includes fundamental commitments in each of the worldview components, and these condition the way in which research toward completing and refining the GSW and the search for a GST* proceeds. In particular, accepting the very concept of a GST* already implies a commitment to certain worldview tenets. Most fundamentally, the GSW outlook is a systems-oriented moderate scientific realism. It is realistic in that it holds that the world has some objective aspects that we can have knowledge of; scientific in that it takes seriously the findings, methods and standards of science; it is moderate in that it acknowledges the limitations and conditionality of our knowledge and our ability to improve it; and it is systems-orientated in that it uses the systems concept to analyse the organization and dynamics of the concrete world.

The philosophical tenets underpinning the GSW were not explicitly developed in the writings of the early general systemists, but appear incidentally in certain works such as von Bertalanffy's classic paper *An Essay on the Relativity of Categories* (von Bertalanffy, 1955). However in 1970s efforts *were* made, notably by Laszlo and Bunge, to develop groups of philosophical statements that would constitute a metaphysical framework for systems science (for example Bunge, 1979; Laszlo, 1972a), and an important overview is given in (Cavallo, 1979).

For present purposes we can summarize the key tenets of GSW using a framework of seven positions. Very briefly, the fundamental philosophical tenets of the GSW are:

T1. *Moderate Epistemological Realism:* We can progressively gain more complete real knowledge of the real world (Bunge, 1973, p. 28; von Bertalanffy, 1955, pp. 258–259);

T2. *Moderate Ontological Realism:* A real concrete world underlies some of our experiences (but experiences can also be distorted or constructed or hallucinated) (Bunge, 1977, p. 16 (M1));

T3. *Broad Naturalism:* Nothing supernaturalistic exists, but concrete phenomena cannot all be reduced to Physics (von Bertalanffy, 1955, pp. 261–262);

T4. *Moderate Systemic Realism:* The concrete world is inherently systemic (but we can also project systemicity onto our experienced world) (Bunge, 1973, pp. 30–31);

T5. *Systemic Universalism:* Every concrete thing (i.e. everything that has causal powers) is always a real system or part of one (Bunge, 1979, pp. 44, 245);

T6. *Moderate Axiological Realism:* Values are largely constructed via cultural processes, but natural systemic processes also influence them (Boulding, 1956b; Rapoport, 1973, p. 247); and

T7. *Moderate Praxeological Realism:* We have the capacity and freedom for uncoerced choices and actions, but our choices and actions can also be conditioned by natural and cultural factors (Rapoport, 1953; von Bertalanffy, 1967, p. 114).

These seven tenets are all metaphysical claims, in that they are about the nature of what exists most fundamentally or about what is inherently possible, but they bear on the full scope of a GSW. Specifically, they have implications for all six of the elements of a worldview as discussed earlier: T1 bears particularly on epistemology, T2 on ontology, T3 on metaphysics, T4 and T5 on cosmology, T6 on axiology and T7 on praxeology.

These tenets are not uncontroversial, but they are consistent with a mainstream position in the philosophy of science, which is a moderate form of scientific realism (Bunge, 2001; Psillos, 1999). For example, these tenets can be mapped onto a similar but narrower framework that Mario Bunge presents as "the philosophical matrix of scientific progress" (Bunge, 2010, pp. 242–245), which in his terminology consists of Scientism (T1), Realism (T2), Materialism (T3), Systemism (T4 and T5), and Humanism (T6 and T7).

Moderate Scientific Realism is not the only metaphysical position in the philosophy of science, but in recent years support for it has been growing in disciplinary areas it is not traditionally associated with, such as the social sciences and management science (Archer, 2013; Mingers, 2014). This trend suggests that the early systemists were ahead of their time in their philosophical commitments.

4.3.4   Arguing from GSW's Tenets to the Potential Existence of GST*

Taken together, the tenets T1–T7 listed above entail not only the existence of a GST*, but moreover that GST* has the kind of potential ascribed to it by the early systemists.

If we assume that a real concrete world exists (T2), and that we can have a scientific model of it (T2 & T3), and that there are real systems in the concrete world (T4), then by implication, there is a scientific theory that models the systemic aspects of the concrete world.

Granted this, if we assume that *all* concrete properties are conditioned by systemic processes (T5), it follows that there is a scientific theory about systemicity that applies everywhere and always.

Hence there exists a GST*.

However, this argument goes beyond a mere existence claim, because if GST* is a theory involving principles that apply everywhere and always, then it has the same ubiquity and utility as general 'Laws of Nature' such as Conservation of Energy and the General Theory of Relativity. Discovering and developing a GST* could thus be of profound significance for science. Not only that, but under the tenets of GSW, GST* would also have implications that go beyond those usually associated with such Laws of Nature, just as the early general systemists proposed.

First, if values are to some degree systemically conditioned in a naturalistic way (T6), then GST* would be relevant to both naturalistic and humanistic concerns. Second, if we have agency and free will (T7), then we can use our knowledge and our values to make a difference to how things turn out, so we can in practice use the insights provided by GST* to change how the world evolves.

These are important inferences, but of course they hinge critically on the validity of the foundational tenets of the GSW. Given the unproven (but not wholly controversial) nature of these tenets, a careful articulation and modern defence of these foundational philosophical assumptions are important outstanding tasks for a contemporary general systems research agenda. In the meantime, we acknowledge that these tenets form a foundational but provisional assumptive framework for General Systemology and the arguments to follow in the rest of this chapter.

4.4   The Potential of the General Systems Worldview (GSW) to Support the Development of GST*

So far, we have shown, on the basis of arguments grounded in the tenets of the GSW, that we can have some confidence that a GST* exists in principle, and that it would be of great practical value to have it. We will now go further, and argue that the GSW can also support the discovery and development of GST*. To develop this

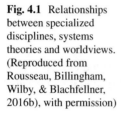

Fig. 4.1 Relationships between specialized disciplines, systems theories and worldviews. (Reproduced from Rousseau, Billingham, Wilby, & Blachfellner, 2016b), with permission)

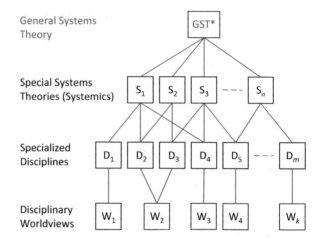

argument we will first discuss an insight into the synergy between GST* and the GSW.

4.4.1 The Relationship of GST* to the Systemics and the Specialised Disciplines

The quotation from von Bertalanffy given earlier makes it clear that he thought that if there were general systems principles, then these would manifest as structures or behaviours that are isomorphic between different kinds of systems. In this sense, identifying and studying such "isomorphies" might present a fruitful approach to discovering the general principles of a GST*, and such an approach has been advocated and actively pursued by Len Troncale (Friendshuh & Troncale, 2012; Troncale, 1978, 1984, 1985, 2009). However, to date the study of specific systemic structures and behaviours has only resulted in more-or-less autonomous theories about specific systemic aspects, for example control theory, network theory, automata theory, hierarchy theory, dissipative structure theory, and so on. Mario Bunge coined the term "Systemics" for the set of these theories that each deal with a specific "isomorphy" (Bunge, 1979, p. 1). Although these isomorphies have not yet been assimilated into a general theory (Francois, 2004, p. 248), it can be said, broadly speaking, that the Systemics pick out 'patterns' that recur across multiple kinds of natural systems (and hence across the specialised disciplines), and that GST* would pick out other kinds of 'patterns' that recur across all the Systemics, as roughly indicated in Fig. 4.1. Here, the individual disciplines are represented as a series of specialised disciplines D_i and the Systemics as a series of systems theories S_i.

As also indicated in Fig. 4.1, the specialized disciplines each have their implicit or explicit worldviews W_i. Some disciplines of course share worldviews, for example Chemistry and Geology, which are both grounded in Scientific Realism and

Physicalism, while others are very different, for example the Social Sciences typically embrace a Constructivist or Postmodern perspective.

4.4.2 GSW as a Counterpart of GST*

The Systemics and GST* are *formal* theories, that is, they contain no information about how the systems they describe are implemented. For example, Communication Systems Theory (Shannon, 1948) describes the functions and limitations of a communication system (for example encoding, signal transmission, detection, noise mitigation, decoding) but does not tell us anything concrete about the many ways in which such components as signal transmitters and receivers might be realized (for example vocal cords, ears, TV antennas). Their lack of ontological commitments guarantees the Systemics' general applicability, but it does raise a puzzle as to why they should be effective in describing real-world phenomena across multiple domains, given that the disciplines in which they apply sometimes have dissonant ontological models. For example, both social systems and mechanical systems exhibit systemic properties such as emergence, synergy and dynamic stability, and yet macro-physical scientists typically assume the existence of an objective reality while social scientists mostly regard reality as a social construction.

The solution to this puzzle was proposed by Ervin Laszlo In his book *Introduction to Systems Philosophy: Toward a New Paradigm of Contemporary Thought* (1972a). Laszlo's argument can be summarized as follows (see Fig. 4.2).

As Laszlo explains it, the existence of specialized disciplines (Physics, Chemistry, Genetics, Sociology etc.) shows that the concrete world is *organized into intelligible domains*. The Systemics, by revealing patterns that recur isomorphically across these domains, cumulatively show that the concrete world is intelligibly organized *as a whole*. This global organization would be reflected in the principles and models of GST*. The existence (in principle) of global organizing principles entails that the concrete world's special domains (as characterized by the specialized disciplines) are contingent expressions or arrangements or projections of a unified underlying intelligibly ordered reality (Laszlo, 1972a, p. 19). In this way Laszlo argued that:

(a) the *existence* (in principle) of GST* implies that there is an intrinsically ordered, and hence unified, reality *underlying* Nature (designated here by the "General Systems Ontology (GSO)" in Fig. 4.2) and

(b) the *content* of GST* provides an abstract model of the systemic nature of this concrete underlying reality (designated here by the "General Systems Metaphysics (GSM)" in Fig. 4.2).

In this light the metaphysical nature of the underlying reality provides the conditions for the manifestation of systemic structures and behaviors in the specialized disciplines, since their phenomena are all grounded in a unified reality that is systemic in nature.

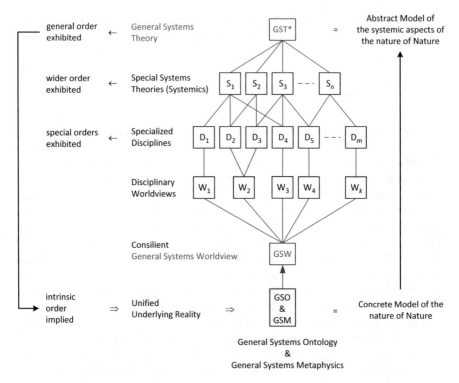

Fig. 4.2 Relationship between GST* and the General Systems Worldview. (Reproduced from (Rousseau et al., 2016b), with permission)

The specialised disciplines all have explicit or implicit worldviews, and these each have an ontological and metaphysical dimension. At present these are not aligned in the way that Laszlo's argument suggests they might be. However, his argument suggests that present-day metaphysical differences between the different worldviews are a historical contingency, and that as science progresses these specialised worldviews will converge in their foundational metaphysical commitments, so that despite their specialised differences they will become *consilient*, reflecting the unity of the underlying reality.

This does not imply that these currently distinct worldviews will collapse into a single 'master' worldview, but it does imply that none of the disciplines will ultimately carry foundational implications that are inherently contradictory to any other's.

4.4.3 The Value of GSW for Developing a GST*

An interesting implication of this argument is that work towards developing the GSW can support the discovery and development process for GST*, in that the two are linked via the metaphysical framework we have called GSM. Via the GSM bridge advances in either GSW or GST* will inform and advance the other. This helps us because although it has been difficult to make progress with developing a GST* the same is not true of the GSW.

The development of a GSW is not dependent on progress towards a GST*, but can proceed on the basis of the findings arising in the specialized disciplines. Many components of the GSW have been worked on by various systems philosophers and others. We already noted earlier the work that has been done towards the metaphysical element of the GSW. Other areas have also been worked on, for example Laszlo included a basic version of a systems cosmology in his *Introduction to Systems Philosophy*, and produced various versions of such models in his later works. (for example Laszlo, 1972a, 1972b, 1994, 1995, 1996, 2004, 2006). Other systems philosophers have also made such contributions (e.g., Bahm, 1967; Boulding, 1985; Bunge, 1979, 2010; Rousseau, 2011; Sutherland, 1973; Vidal, 2013), and the project of building up a GSW is healthy if still radically incomplete, in a fragmented state, and in the custody of a rather small group of researchers.

However, to date the objective of this work was not to find the principles for a GST*, but simply to find a coherent philosophical perspective on the basis of scientific knowledge and the systems paradigm. Some prominent researchers in this area do not even believe that a GST* can exist (e.g. Bunge, 2014, p. 11), so it is certainly not the case that researchers are always on the lookout for general systems principles. This represents a missed opportunity for advancing GST*, as the argument we presented above shows. We can rectify this omission by a change of attitude in developing a GSW, in which we are particularly alert to the potential for discovering general systems principles when analyzing the metaphysical implications of the specialized sciences.

This work can be facilitated by taking a more systematic approach, in which we summarize and compare the worldviews of the specialized disciplines in a consistent way. This could be done by first constructing a systems-oriented model of the structure and scope of a worldview, and using this as a template for recording the basic commitments of the specialized worldviews. This will help us to identify common foundations but also metaphysical conflicts between worldviews. The former would represent the core of an emerging integrated GSW, and the latter could identify questions for investigation using a systems approach. The "GSM-argument" discussed earlier indicates that an ultimate metaphysical pluralism is not realistically sustainable, and therefore the places where it occurs would provide rich opportunities for scientific discovery.

As the "core GSW" emerges from this comparison exercise, so would we develop better clarity about the metaphysical foundation that links GSW and GST*. The richness of the material available in this area of work is immense. The opportunity for discovering general systems principles when working systematically with the basic findings of all the disciplines must be very substantial, and much greater than when trying to abstract such principles from the study of a relatively small number of isomorphies.[2]

---

[2] Len Troncale has claimed that about a hundred isomorphies have been identified (Friendshuh & Troncale, 2012; Troncale, 1978), but it is unclear to us whether his list employs the term "isomorphy" as narrowly as meant by von Bertalanffy.

If it is true that the dynamics of all the structures evolving throughout nature are exemplifying underlying general systems principles, and all the kinds of systems we find in nature behave in ways consistent with general systems principles, then these principles can be expected to 'shine through' the data describing the world, if the data is organized in an appropriate systemic way.

What we are seeking in constructing GSW in a systemic way is not merely a *taxonomy*, organizing the data in line with a set of empirical criteria (Bailey, 1994), but a representative *typology*, a classification according to concepts that 'carve at the joints' of reality, or at least that part of reality that is represented by the body of scientific knowledge (Sider, 2011). Inverting Boulding's famous dictum, if Systems Philosophy can find the joints of the body of science, then it can be opened up to reveal the skeleton on which its integrity depends, GST*.

The development of such a worldview comparison framework is thus an important initial step towards a new and promising strategy for accelerating progress towards GST*, and should be added to the research objectives of a contemporary research agenda for General Systemology.

4.5   The Potential Value of the Synergy Between GST* and GSW

We have argued that Systems Philosophy can help develop a GSW, and that this will not only aid the development of GST*, but that progress with GST* will in turn inform and enrich the development of the GSW. However, it is important to emphasise that developing a GST* and/or a GSW are not the most important goals at stake here, but rather that they are valuable for what they each enable.

A GST* would provide a framework from which we can discover, in a principled way, kinds of systemic structures and systemic behaviours unanticipated by contemporary science. This is important for it heralds the discovery of new ways to understand, design, engineer or govern systems. A GSW, on the other hand, embodies our best understanding of the nature, state, and potential of the world as a total system, providing us with a framework for discussing questions of ultimate concern. Moreover, using the GSW framework to compare and analyze worldviews we can identify opportunities for systems research that can deepen or extend our fundamental insights. Taken together, the mechanisms newly identified in the concrete world due to the development of GST*, and the potentials in the concrete world newly identified by developing GSW, can open up significant new avenues of systemic intervention.

In Fig. 4.3 we present this view of General Systemology's scope in a schematic way. We have here used the same colour scheme as we did for the "AKG Model" of a discipline we presented in Chap. 2, and used blue for components of the Knowledge

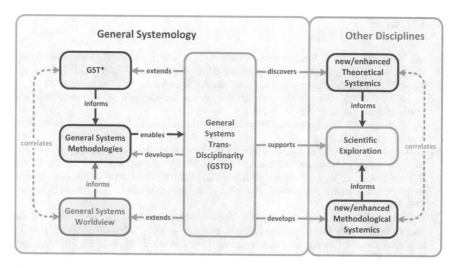

Fig. 4.3 General Systems Transdisciplinarity as the activity scope of General Systemology. (Reproduced from (Rousseau et al., 2016b), with permission)

Base, orange for components of the Guidance Framework, and green for components of the Activity Scope.

This framework heralds a new era of General Systems Transdisciplinarity, in which we use GST* and GSW as reference baselines for methods of doing fundamental research towards new Systemics and new fundamental insights, and use these advances to develop methods for future waves of systemic intervention towards building the 'better world' the founders of the general systems movement envisioned. Such an extended version of General Systemology would realize the General Systems Transdisciplinarity that our present world needs even more urgently than it did at the founding of the general systems movement.

4.6 Summary

In this chapter we discussed the historical debate about the potential existence of a GST*, and built a case for its potential existence and practical value. We developed a model of the "General Systems Worldview" (GSW), and related it to the general model of a worldview we started to develop in Chap. 3. We showed that the assumptions of the GSW entail the potential existence of a practically valuable GST. We discussed how the relationship between the GSW and GST* suggests that the GSW can guide us to the discovery of general systems principles, and argued that both the GSW and GST are needed for the development of GSTD.

References

Adams, K. M., Hester, P. T., Bradley, J. M., Meyers, T. J., & Keating, C. B. (2014). Systems theory as the foundation for understanding systems. *Systems Engineering, 17*(1), 112–123.

Archer, M. S. (2013). *Social morphogenesis*. Dordrecht: Springer.

Bahm, A. J. (1967). Organicism: The philosophy of interdependence. *International The Philosophical Quarterly, 7*(2), 251–284.

Bailey, K. D. (1994). *Typologies and taxonomies: An introduction to classification techniques*. Thousand Oaks, CA: Sage.

Billingham, J. (2014a). *GST as a route to new Systemics*. Presented at the 22nd European Meeting on Cybernetics and Systems Research (EMCSR 2014), Vienna, Austria. In J. M. Wilby, S. Blachfellner, & W. Hofkirchner (Eds.), EMCSR 2014: Civilisation at the crossroads – Response and responsibility of the Systems Sciences, Book of abstracts (pp. 435–442). Vienna: EMCSR.

Billingham, J. (2014b). *In search of GST*. Position paper for the 17th conversation of the International Federation for Systems Research on the subject of 'Philosophical foundations for the modern systems movement', St. Magdalena, Linz, Austria, 27 April–2 May 2014, pp. 1–4.

Boulding, K. E. (1956a). General systems theory – The skeleton of science. *Management Science, 2*(3), 197–208.

Boulding, K. E. (1956b). *The image: Knowledge in life and society*. Ann Arbor, MI: The University of Michigan Press.

Boulding, K. E. (1985). *The world as a total system*. Beverly Hills, CA: Sage.

Bunge, M. (1973). How do realism, materialism, and dialectics fare in contemporary science? Reproduced in M. Maher (Ed.). (2001) *Scientific realism* (pp. 27–41). Amherst: Prometheus. Page references in the present paper refer to the reproduction. In *Method, model and matter* (pp. 169–185). Dordrecht: Reidel.

Bunge, M. (1977). *Ontology I: The furniture of the world*. Dordrecht: Reidel.

Bunge, M. (1979). *Ontology II: A world of systems*. Dordrecht: Reidel.

Bunge, M. (2001). *Scientific realism: Selected essays of Mario Bunge*. (M. Mahner, Ed.). Prometheus Books.

Bunge, M. (2010). *Matter and mind: A philosophical inquiry*. New York: Springer.

Bunge, M. (2014). Big questions come in bundles, hence they should be tackled systemically. *Systema, 2*(2), 4–13.

Cavallo, R. E. (1979). Systems research movement: Characteristics, accomplishments, and current developments. *General Systems Bulletin, 9*(3), 1–131.

Checkland, P. (1993). *Systems thinking, systems practice*. New York: Wiley.

Dubrovsky, V. (2004). Toward system principles: General system theory and the alternative approach. *Systems Research and Behavioral Science, 21*(2), 109–122.

Elohim, J.-L. (2000). *A General systems WELTANSCHAUUNG (Worldview)*. Retrieved January 20, 2016, from http://www.isss.org/weltansc.htm

Flood, R. L., & Robinson, S. A. (1989). Whatever happened to general systems theory? In R. L. Flood, M. C. Jackson, & P. Keys (Eds.), *Systems prospects* (pp. 61–66). New York: Plenum.

Francois, C. (Ed.). (2004). *International encyclopedia of systems and cybernetics*. Munich: Saur Verlag.

Francois, C. (2007). *Who knows what general systems theory is?* Retrieved January 31, 2014, from http://isss.org/projects/who_knows_what_general_systems_theory_is

Friendshuh, L., & Troncale, L. R. (2012, July 15–20). *Identifying fundamental systems processes for a general theory of systems*. In Proceedings of the 56th annual conference, International Society for the Systems Sciences (ISSS), San Jose State University, 23 pp.

Georgiou, I. (2006). *Thinking through systems thinking*. New York: Routledge.

Hiebert, P. G. (2008). *Transforming worldviews: An anthropological understanding of how people change*. Grand Rapids, MI: Baker Academic.

Hofkirchner, W. (2013). *Emergent information: A unified theory of information framework.* London: World Scientific.

Kant, I., & Gregor, M. J. (1987). *Critique of judgment* (W. S. Pluhar, Trans.). Indianapolis: Hackett.

Laszlo, E. (1972a). *Introduction to systems philosophy: Toward a new paradigm of contemporary thought.* New York: Gordon & Breach.

Laszlo, E. (1972b). *The systems view of the world: The natural philosophy of the new developments in the sciences.* New York: George Braziller.

Laszlo, E. (1994). From GUTs to GETs: Prospects for a unified evolution theory. *World Futures, 42*(3), 233–239.

Laszlo, E. (1995). *The interconnected universe: Conceptual foundations of transdisciplinary unified theory.* London: World Scientific Publishing.

Laszlo, E. (1996). *The systems view of the world: A holistic vision for our time.* Cresskill, NJ: Hampton Press.

Laszlo, E. (2004). *Science and the Akashic field: An integral theory of everything.* Rochester, VT: Inner Traditions.

Laszlo, E. (2006). *Science and the reenchantment of the cosmos: The rise of the integral vision of reality.* Rochester, VT: Inner Traditions.

Mingers, J. (2014). *Systems thinking, critical realism and philosophy: A confluence of ideas.* New York: Routledge.

Naugle, D. (2002). *Worldview: The history of a concept.* Cambridge: Eerdmans.

Pouvreau, D. (2013). The project of "general systemology" instigated by Ludwig von Bertalanffy: Genealogy, genesis, reception and advancement. *Kybernetes, 42*(6), 851–868.

Pouvreau, D. (2014). On the history of Ludwig von Bertalanffy's "general systemology", and on its relationship to cybernetics – Part II: Contexts and developments of the systemological hermeneutics instigated by von Bertalanffy. *International Journal of General Systems, 43*(2), 172–245.

Psillos, S. (1999). *Scientific realism: How science tracks truth.* London: Routledge.

Rapoport, A. (1953). *Operational philosophy: Integrating knowledge and action.* San Francisco: International Society for General Semantics.

Rapoport, A. (1973). Review of Laszló: The systems view of the world. *General Systems, XVIII,* 189–190.

Rousseau, D. (2011). *Minds, souls and nature: A systems-philosophical analysis of the mind-body relationship in the light of near-death experiences* (PhD thesis). University of Wales Trinity Saint David, Lampeter, Wales, UK.

Rousseau, D., Billingham, J., Wilby, J. M., & Blachfellner, S. (2016a). In search of general systems theory. *Systema Special Issue – General Systems Transdisciplinarity, 4*(1), 76–92.

Rousseau, D., Billingham, J., Wilby, J. M., & Blachfellner, S. (2016b). The synergy between general systems theory and the general systems worldview. *Systema Special Issue – General Systems Transdisciplinarity, 4*(1), 61–75.

Shannon, C. E. (1948). A Mathematical Theory of Communication. *Bell Systems Technical Journal, 27*(Part 1), 379–423.

Sider, T. (2011). *Writing the book of the world.* Oxford: Oxford University Press.

Simon, H. A. (1996). Alternative views of complexity. In *The sciences of the artificial* (pp. 169–181). Cambridge, MA: MIT Press.

Sire, J. W. (2004). *Naming the elephant: Worldview as a concept.* Downers Grove, IL: IVP Academic.

Sutherland, J. W. (1973). *A general systems philosophy for the social and behavioral sciences.* New York: George Braziller.

Troncale, L. R. (1978). Linkage propositions between fifty principal systems concepts. In G. J. Klir (Ed.), *Applied general systems research* (pp. 29–52). New York: Plenum Press.

Troncale, L. R. (1984). What would a general systems theory look like if I bumped into it? *General Systems Bulletin, 14*(3), 7–10.

Troncale, L. R. (1985). The future of general systems research: Obstacles, potentials, case studies. *Systems Research, 2*(1), 43–84.

Troncale, L. R. (2009). Revisited: The future of general systems research: Update on obstacles, potentials, case studies. *Systems Research and Behavioral Science, 26*(5), 553–561.

Vidal, C. (2013, January 5). *The beginning and the end: The meaning of life in a cosmological perspective*. Free University Brussels. Retrieved from http://arxiv.org/abs/1301.1648

von Bertalanffy, L. (1934). Wandlungen des biologischen Denkens. *Neue Jahrbücher Für Wissenschaft Und Jugendbildung, 10*, 339–366.

von Bertalanffy, L. (1955). An essay on the relativity of categories. *Philosophy of Science, 22*(4), 243–263.

von Bertalanffy, L. (1967). *Robots, men and minds*. New York: Braziller.

Wilby, J. M., Rousseau, D., Midgley, G., Drack, M., Billingham, J., & Zimmermann, R. (2015). Philosophical foundations for the modern systems movement. In M. Edson, G. Metcalf, G. Chroust, N. Nguyen, & S. Blachfellner (Eds.), *Systems thinking: New directions in theory, practice and application*, proceedings of the 17th conversation of the International Federation for Systems Research, St. Magdalena, Linz, Austria, 27 April–2 May 2014 (pp. 32–42). Linz: SEA-Publications, Johannes Kepler University.

Chapter 5
The Knowledge Base of General Systemology

Abstract The search for a foundational general systems theory (GST*) formally became a scientific enterprise with the founding of the Society for the Advancement of General Systems Theory in 1954. Many scientific advances have been made towards a GST*, but GST* is still incomplete and there is a rich ongoing debate about the nature, structure and value of GST*. In this chapter we argue that the general theory of a discipline has a generic structure, which can be inferred by attending to the process by which disciplines build up their knowledge base. We develop a model of this generic structure and then use it to envision the structure and scope of GST*. This provides a principled baseline for assessing the developmental status of GST*, planning work towards its completion, and defending the potential value of GST*.

Keywords General Systems Theory · GST · GST* · General Systems
Transdisciplinarity · GSTD · Boulding · AKG model

5.1 Introduction

The existence, in principle, of a GST* was first suggested about a hundred years ago (Bogdanov, 1913; von Bertalanffy, 1932), but the search for it only became an organised scientific enterprise with the founding of the Society for the Advancement of General Systems Theory in 1954, which lives on today as the International Society for the Systems Sciences (ISSS)).

Much progress has been made in developing the theories and methods of systems science since the 1950s, but progress towards discovering the principles of a GST* has been limited, and until recently there remained a considerable lack of clarity around what a GST might look like and how it might be discovered (Francois, 2006).

For a long time our best impressions of what a GST* would be like were due to von Bertalanffy and Kenneth Boulding. From von Bertalanffy we have the idea of a theory incorporating principles that underpin the development of isomorphic systemic structures and processes in systems (see Chap. 1, Sect. 1.2), and he also identified one candidate for such a principle, namely the proposition that there are no closed systems in nature, i.e. all naturalistic systems are open (to various degrees) to cross-boundary flows of energy, matter and information (von Bertalanffy, 1950). Unfortunately von Bertalanffy did not offer any guidance about how such principles might be deduced from the study of isomorphies, and we arguably so far have only one other clear-cut candidate, namely Herbert Simon's proposition that viable complex systems are near-decomposable hierarchies (Simon, 1962).[1] Lack of progress with identifying such principles led Kenneth Boulding to suggest that perhaps GST* might in fact only be a theory about isomorphies and utility rather than deeper principles (Boulding, 1956, p. 198).[2] However, Boulding also held out a higher hope:

> At a low level of ambition but with a high degree of confidence [GST*] aims to point out similarities in the theoretical constructions of different disciplines, where these exist, and to develop theoretical models having applicability to at least two different fields of study. At a higher level of ambition, but with perhaps a lower degree of confidence it hopes to develop something like a "spectrum" of theories—a system of systems which may perform the function of a "gestalt" in theoretical construction (Boulding, 1956, p. 198).

This passage is from Boulding's classic paper *General System Theory – The Skeleton of Science*, in which he introduced the notion of (what we call) GST* as a spectrum of systems theories representable as a kind of "periodic table of systems". This was an inspiring notion, given the impact of the discovery of the periodic table in Chemistry. In Chemistry the fragmentary initial table produced by Mendeleev immediately made it possible to predict the existence of undiscovered elements in nature, and to characterise them to some degree so that effective empirical searches for them could be undertaken. It also cast valid suspicion on certain accepted empirical findings, inspiring more refined research that corrected these errors. If this situation could be obtained for *systems* the new 'periodic table' might enable the discovery of hitherto unknown and unsuspected kinds of systemic structures, behaviours or capacities existing in nature, opening the way for more effective systemic methodologies.

In *Skeleton* Boulding sketched out a now well-known hierarchy of system types (see Fig. 5.1) that could be taken as the first contribution towards such a 'periodic table', but it has proven difficult to build on this. A contributing factor may be that there are multiple ways of interpreting what the hierarchy is ordering over (Billingham, 2014a; Mingers, 1997; Wilby, 2006). However, the vision remains

---

[1] George Mobus and Michael Kalton recently proposed a useful list of heuristic "principles of systems science" (Mobus & Kalton, 2014, pp. 17–30), that could form a useful starting point for developing general systems principles that can be explicitly connected to the manifestation of systemic isomorphies.

[2] This view is defended by Mario Bunge, who proposes that GST* does not exist as a distinct theory but only as a collection of specialised theories about each of the kinds of systemic structures and processes (Bunge, 1979, p. 1; 2014, p. 8). A similar but more refined position has been pursued by Len Troncale, who *does* takes GST* to be distinct from theories about the isomorphies, but sees GST* as a model of the linkages between the isomorphies rather than the principles underlying them (Friendshuh & Troncale, 2012; Troncale, 1978, 1986, 1988). According to Troncale, the isomorphies are not merely correlations between aspects of models of kinds of systems, but isomorphies are objectively real and are the causes of the manifestations of systemic structures and processes (Troncale, 1988, p. 17).

Fig. 5.1 The systems
hierarchy from Boulding's
'Skeleton of Science'
(1956). (Reproduced from
(Rousseau, Billingham,
Wilby, & Blachfellner,
2016), with permission)

Level	System Type
9	Transcendental systems
8	Social organizations
7	Human
6	Animal
5	Genetic-societal level
4	Open systems
3	Cybernetic system
2	Clockworks
1	Frameworks

inspiring – for example, one of the charter ambitions of the Systems Science Working Group (SSWG) (founded in 2011) in the International Council on Systems Engineering (INCOSE) is "Formulation of a Periodic Table of Systems" (Martin, 2011).

In recent work one of the present authors (Billingham) has, on the basis of the historical development of general theories in science, reinterpreted and expanded Boulding's metaphor (Billingham, 2014a, 2014b; 2015). First, it was argued that Boulding was attempting to specify a spectrum of kinds of systemic behavioural capacities, but lacked the terminology to specify his vision clearly. Such an interpretation makes a much clearer connection with the notion of a periodic table such as Chemistry's, which can be interpreted as a typology classifying behavioural potentials. Second, it was argued that formulating such a 'gestalt' depends on also developing other general theoretical components not described in *Skeleton*, such as abstract models of generic entities and theoretical mechanisms underpinning the generation of empirical instances. For example, in Chemistry the periodic table of elements is not only an *empirical* model based on an ordering of observed properties of kinds of elemental substances, but it can also be understood as following logically from the application of instantiation rules (such as the sequence in which electron orbitals are populated) to the general model of an atom, and as underpinning our theoretical understanding of how atoms combine into more complex substances. A general theory is therefore more adequately conceived of as an interlocking array of general principles, models and theories, rather than as a single construct such as the periodic table, as illustrated for the case of Chemistry in Fig. 5.2.

As Billingham pointed out, such theoretical construction has independent value beyond the discovery of unknown 'elements' indicated by gaps in the empirically-determined table. Certain kinds of chemical elements, such as the noble gases and radioactive isotopes could not be predicted from a fragmentary empirical table of elements, but required further breakthroughs in fundamental Chemistry, such as making the distinction between elements, compounds and mixtures, and the

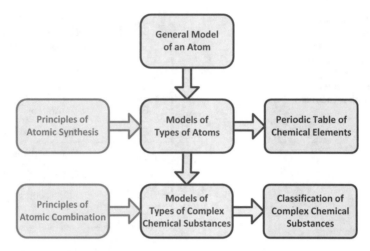

Fig. 5.2 A model of general theory in Chemistry. (Reproduced from (Rousseau, Billingham, et al., 2016), with permission)

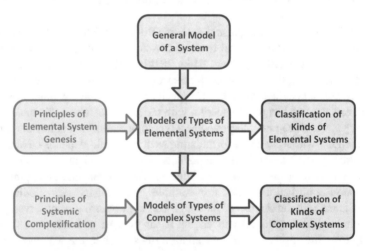

Fig. 5.3 A model of GST* based on general theory in Chemistry. (Reproduced from (Rousseau, Billingham, et al., 2016), with permission)

development of more sophisticated generic atomic models (for example the wave mechanics models based on the work of Schrödinger). Progress towards a fully-fledged GST* is therefore likely to depend as much on theoretical advances towards general models and general principles as on exploration of empirical kinds of systemic behaviours. Accordingly, a GST* could be envisioned more accurately as having a structure as shown in Fig. 5.3.

This model provides a more comprehensive vision of what a GST* would encompass, and could be a suitable basis for new strategies to develop the requisite components. The nature of the systems 'elements' is still undetermined, but Billingham postulated that these would be minimal systemic constructs each adding a unique archetypal behaviour. However, as Billingham noted, this extension of Boulding's metaphor should still be regarded with some caution, as it is based entirely on an analogy with a purely physical system, whereas a GST* would have to encompass all kinds of concrete[3] systems, and thus also ones with properties that are still mysterious from the perspective of physical science for example agency, will, intentionality, volition, consciousness, creativity, and self-awareness (Gillett & Loewer, 2001; Rousseau, 2011). It is unclear to what extent this metaphorical analogy would have to be modified/extended to accommodate this wider scope. Other questions could also be raised, for example whether it is sufficiently general (i.e. would a different structure be proposed if another discipline's general theory was used as a basis for the metaphor?) and whether it is complete (i.e. would an analogy based on the general theory of a more complex discipline (for example Biology or Sociology) reveal additional components?).

5.2   A General Perspective on General Theories

A key aim of the research reported in the present chapter is to remove this kind of ambiguity from the debate about the nature of GST and establish a more principled foundation for envisioning the nature of GST. The GSTD team realised that the mentioned concerns could be addressed by employing the generic model of the structure of a scientific discipline, called "the AKG model", which was presented in Chap. 2. This model shows that a discipline is a kind of system with dynamic interactions between its Activity Scope, Knowledge Base and Guidance Framework.

Our idea was that by using a generic model it would be possible to generalize (and if appropriate refine or extend) Billingham's model without relying exclusively on a model of the general theory of a specific discipline. Our challenge in executing on this idea was that we had not in previous work expanded the knowledge base component of the AKG model in sufficient detail for it to be immediately useful in this specific context. However, we thought that we could readily expand this model component by taking note of generic lessons we learnt from the history of science about how disciplinary knowledge is built up, and tying it in with general ideas about knowledge bases explored in Chap. 2. In Chap. 2, we argued that the AKG model shows that all scientific disciplines (and disciplinary fields) can be modelled as having both a similar structure and similar dynamics in their development, and that this applies also to Systemology, even though it is a transdiscipline. In Chap. 2 we also argue that each discipline has a unifying theory, and that this is a "general theory" in that it applies always and everywhere within its discipline. We argued

---

[3] *Concrete* systems are systems with causal powers. GST* was conceived as a general theory over natural systems, hence the specification of its scope as embracing concrete systems. However, it may have significance for abstract systems too.

that for Systemology that unifying theory would be GST*. On the basis that this model shows disciplines to have a generic structure and generic dynamics, we suggest that the general theories of all disciplines have a similar structure to each other too, and are also developed in similar ways. Consequently, we would therefore suggest that GST*, as the general theory of Systemology, will have a similar structure (and developmental pathway) to other general theories in other disciplines (generically, not only compared to Chemistry). In this chapter we will therefore expand the generic model of the knowledge base of a discipline, to show the generic structure of the general theory component (and its generic context), and from this propose where to look, and what to model, as we search for a GST*. In this way we hope to present a conception of the scope and structure of a GST* that can guide research towards its development in a more systematic way than has been available previously.

For present purposes we need not expand the complete model of the knowledge base of a discipline. As discussed in relation to the AKG model in Chap. 2, the knowledge base of a discipline consists of data comprising observations and findings, theories comprising special, general and hybrid theories, and methodologies comprising heuristic, special, general and hybrid methodologies. Our immediate concern is with general theories and their precursors, so we will for now ignore hybrid theories and all kinds of methodologies. However, we will include the guidance framework elements that condition the activity by which the knowledge base is built up, as these directly support the development of the general theories. For completeness we will briefly cover the missing aspects later on in Sect. 5.6, and describe the completed model in Sect. 5.7 at the end of the present chapter.

Our strategy for developing the expanded model of a disciplinary knowledge base is to draw on the history and philosophy of science, by following the stages through which disciplinary activity builds up its knowledge base and guidance framework. We observe that scientific frameworks and core theories are built up cumulatively as scientists (and scientific philosophers) try to answer (or improve answers to) a structured series of generic questions. All these questions can be worked on in parallel, and the answers to each cross-inform the work on others, but overall being able to make good progress with any one is dependent on the progress that has already been made with prior ones. For ease of reference we summarize these questions in Fig. 5.4, before discussing them in more detail in Sect. 5.3. Each question motivates activity relating to a certain kind of disciplinary content, which we will label for convenience of reference. These terms are either used in conventional ways or in ways that generalize their conventional meanings.

Answering Q1 and Q2 produces essential precursors to knowledge generation by setting out the empirical boundary[4] and the technical vocabulary for the investigation. The scope of these is conditioned by worldviews, which can be made explicit by answering Q3. In terms of the AKG Model Q1–Q3 represent components of the discipline's Guidance Framework.

---

[4]"Empirical" is defined in the OED as "based on, concerned with, or verifiable by observation or experience rather than theory or pure logic".

Questions	Content Type	Content Category	AKG Element
1	What qualifies something as a subject entity for the discipline?	Empirical identity criteria for subject entities	Empirical domain boundaries	**Guidance Framework**
2	How can we describe the subject entities?	Technical terms and definitions	Subject terminology	
3	Why do we limit the scope of the discipline as we do?	Perspectives and narratives about knowledge, nature, life and self	Worldviews	
4	What are the subject entities like?	Descriptions of observable features of empirical entities	Morphology	**Data**
5	How do they work?	Studies on the processes that produce/ sustain specific morphological features	Morphodynamics	**Special Theories**
6	How do they come about?	Studies on the intrinsic nature (natural kind) and intrinsic dynamics of the subject entities	Morphogenetics	**General Theories**

Fig. 5.4 Key questions in the build-up a Disciplinary Knowledge Base. (Reproduced from (Rousseau, Billingham, et al., 2016), with permission)

This framing regulates and enables the building of the discipline's Knowledge Base. The foundational element of this is the collection and classification of empirical data (Q3). Data represents pre-theoretical knowledge that underpins scientific theory development, and it documents observable features of the subject entities. We will refer to this study area as "morphology".[5] Data enables theory development, and this commences with activity towards developing specialized explanatory theories about the functions of specific entity features and the particular processes that underlie them (Q4). We will refer to this area of study as "morphodynamics".[6] Data and specialized knowledge sets the stage for work on a natural next question, namely how the subject entities come about (Q6). We will refer to this area of study as "morphogenetics".[7] Q6 is pragmatically addressed via four subsidiary questions (that are once again most easily addressed in a cumulative way), namely: *how do the simplest subject entities come about?*, *how do complex entities come about?* and

---

[5] We adopt this technical term from its usage in Biology. It derives from the ancient Greek μορφή, *morphé*, meaning "form", and λόγος, *lógos*, meaning "study, research". In the present context we interpret the notion "form" very widely, to include all aspects of appearances for example shape, structure, composition, colour, functions, behaviours, properties, powers, capacities etc.

[6] From the ancient Greek μορφή, *morphé*, meaning "form", and from the Greek δυναμικός *dynamikos* "powerful", from δύναμις *dynamis* "power".

[7] From the ancient Greek μορφή, *morphé*, meaning "form", and γενετικός, *genetikos*, meaning "genitive"/"generative", which in turn derives from γένεσις *genesis* meaning "origin". Our usage generalizes the application of the term morphogenetics beyond its current use in biology, where it refers only to the study of the development of normal *organic* form (Merriam-Webster, n.d.).

why do certain kinds of entities or entity designs not arise or persist? The answers to Q6-type questions describe and theorize over factors relevant to all subject entities, and are therefore contributions to the general theories of the discipline. Being common ground for the discipline these theories provide scientific foundations for the unity of the discipline.

Although strong progress with any of these questions typically requires strong progress with 'earlier' questions in this series, it is of course also the case that progress with 'later' questions can provide insights that trigger significant revisions of 'earlier' work, so that this build-up of knowledge is more like a maturing system than a linear growth process.

This 'feedback' loop is particularly evident in relation to general theories. Although general theories are concerned with foundational aspects of the discipline, their development requires much prior progress of specialized kinds, and hence scientifically significant general theories typically arrive late in the life cycle of a discipline. However once they begin to appear they can trigger significant new work and important advances in specialized theories, which in turn can enable new advances in general theory development. They can even cause revision of the domain boundaries, as happened in the separation of Chemistry from Alchemy and Astronomy from Astrology.

We will now briefly explore the role and ramifications of these questions in turn, and this discussion will be followed by a summarizing table (Fig. 5.5) and overview diagrams (Figs. 5.6 and 5.7).

5.3 Core Questions for the Development of a Scientific Knowledge Base

5.3.1 Guidance Framework Questions

The activity of building up a disciplinary Knowledge Base is conditioned and enabled by three prerequisite elements that form part of the disciplinary 'Guidance Framework', namely an empirical definition of the discipline's subject matter, a specification of descriptive technical terms, and worldviews.

5.3.1.1 Q1: What Qualifies Something as a Subject Entity for the Discipline?

The key prerequisite for theory building is data collection, and for this we need to establish empirical boundaries for disciplinary inquiry. These limits are set by answering Q1, thus providing.

a subject definition for the discipline. The definition serves as the initial unifying framework of the discipline, even if only of an administrative or political kind.

#	Primary Questions	Content Type	Content Category	Secondary Questions	Content Subtype	Content Subcategory	Content Example(s)	AKG Elements
1	What qualifies something as a subject for the discipline?	Empirical identity criteria for subject entities	Empirical domain boundaries	What qualifies something as a subject for the discipline?	Empirical identity criteria for subject entities	Empirical domain boundaries	Biology as the study of entities that ingest, metabolise and reproduce	Guidance Framework
2	How can we describe the subject entities?	Vocabulary of technical terms & definitions	Subject terminology	How can we describe the subject entities?	Vocabulary of technical terms & definitions	Subject terminology	Classification key in Biology	
3	Why limit the scope of subject entities and terms as we do?	Perspectives on Knowledge, Nature, Life and Self	Worldviews	Why limit the scope of entities and terms as we do?	Perspectives on knowledge, Nature, Life and Self	Worldviews	Perspectivism, critical realism, constructivism	Data
4	What are subject entities like?	Descriptions of features of empirical subject entities in terms of forms, structures, functions, behaviours, etc.	Morphology	What are the observable features of subject entities?	Entity feature descriptions	Annotations	"Birds have feathers, beak, no teeth, lay hard-shelled eggs"	
4				How are the empirical kinds related?	Classification via cluster analysis of empirical features	Taxonomy	Linnaean System in Biology	
5	How do they work?	Studies on the processes that produce/sustain specific morphological features	Morphodynamics	How do they work?	Studies on the processes that produce/sustain specific morphological features	Morphodynamics	E.g. In biology theories about reproductive & digestive processes	Special Theories
6	How do they come about?	Studies on the origins and life-stages of entity kinds	Morphogenetics					
6.1	How do the simplest individuals come about?	Studies on the intrinsic nature (natural kind) of the subject entities	Elemental Morphogenetics	How do the simplest subject entities come about?	Theories about the nature of the simplest subject entities	Protogenetics	Atom Theory in chemistry; Cell biology in biology	
6.1				What types of elemental entities are possible?	Models based on dependencies between elemental features	Elemental ontology	Types of atoms; Types of cells	
6.1				How are the elemental types related to each other?	Logical classification of elemental kinds	Elemental typology	Standard model in physics; Periodic table in chemistry;	
6.2	How do complex individuals come about?	Studies on how individuals are generated and how they develop	Developmental Morphogenetics	How do the mature complex entities develop from simpler beginnings?	Theories about the processes that govern development of complex forms	Ontogenetics	Embryology in biology; Stellar life cycle in astronomy	General Theories
6.2				What types of complex entities are possible?	Models based on dependencies between complex entity features	General ontology	Types of chemical compounds; Types of complex animals	
6.2				How are the complex kinds related to each other?	Logical classification of complex kinds	General typology	Phylogenetic trees in biology; Hubble galaxy sequence	
6.3	How does the variety of complex individuals come about?	Studies on how new kinds of individuals arise from existing kinds	Evolutionary Morphogenetics	How do new kinds of subject entities come about?	Theories about evolutionary mechanisms	Phylogenetics	Theory of evolution by variation and natural selection in biology	
6.4	Why do some entity types or design patterns not arise or not persist?	Studies on how the architectures of entities are/become optimized for their operational contexts	Axiological Morphogenetics	Why are they not configured in functionally equivalent ways?	Theories about how entity designs support resilience in different environments	Axionetics	Epigenetics in biology; Value-driven systems eng	
6.4				What types of wholes are resilient?	Models based on resilience features of wholes	Ontology of wholes	Types of ecosystems	
6.4				How are kinds of wholes related to each other?	Logical classification of types of wholes	Typology of wholes	An ecosystem panarchy	

Fig. 5.5 Tabular overview of the cumulative development of a disciplinary knowledge base

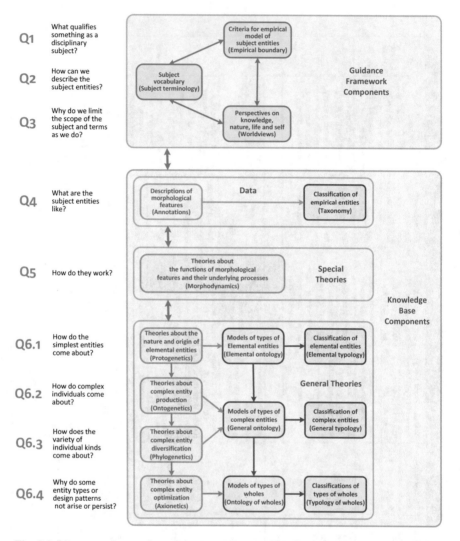

Fig. 5.6 Diagrammatic overview of the cumulative development of a disciplinary Knowledge Base. (Reproduced from (Rousseau, Billingham, et al., 2016), with permission)

Typically, this definition provides empirical criteria for identifying entities of interest, for example in biology the subject entities are 'organisms', and they might be identified as entities that jointly have capacities such as respiration, metabolism, reproduction, growth, self-repair, and homeostasis. The subject entity definition forms part of the "domain definition" of the discipline included in the disciplinary guidance framework as described in Rousseau, Wilby, et al. (2016). This wider definition includes the objectives of the discipline (specifying why the subject entities are studied) and the stance of the discipline (such as specifying limits on how studies might be done for example using ethical criteria).

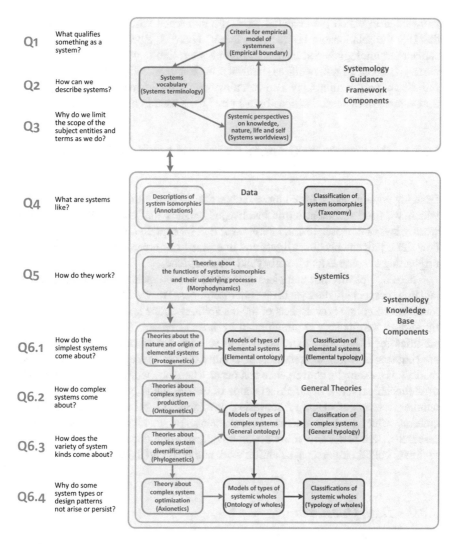

Fig. 5.7 Diagrammatic overview of the cumulative development of Systemology's Knowledge Base. (Reproduced from (Rousseau, Billingham, et al., 2016), with permission)

5.3.1.2   Q2: How Can We Describe the Subject Entities?

Data collection entails making descriptions, and for this a precise vocabulary is needed. A vocabulary associated with a field of study is called a "terminology". To answer Q2 we only need a subset of the discipline's complete terminology, which we will call the "subject terminology". The subject terminology provides standardised terms for characterising disciplinary subject entities in terms of standardised concepts. In the context of scientific theorizing the subject terminology represents a controlled vocabulary that provides "formal representations of areas of knowledge in which the essential terms are combined with structuring rules that describe the relationship between the terms" (Bard & Rhee, 2004, p. 213).[8] The subject terminology is part of the "disciplinary terminology" included in the disciplinary Guidance Framework as described in Rousseau, Wilby, et al. (2016). The full disciplinary terminology also contains other domain-specific technical terms, for example ones needed to specify tasks and tools used in methodologies.

5.3.1.3   Q3: Why Do We Constrain the Scope of Inquiry and Terminology as We Do?

Drawing boundaries for the empirical scope the discipline and defining concepts for which we need technical terms involve judgements that are conditioned by the worldviews of the researchers. Making worldviews explicit thus helps to make clear how these judgements about limits and interpretations are grounded in judgements originating to some extent from beyond the discipline.

There is typically not complete agreement in the discipline about boundaries, and this is especially the case in nascent disciplines. The worldviews of disciplinary members condition how aspects of the disciplinary scope and some of the terms in the subject terminology are interpreted *by those members*. This creates richer possibilities for exploratory activities within science, but as progress is made some convergence of worldviews may occur within a discipline (depending on the subject matter). For example, convergence can occur where one worldview's interpretation of the technical terms or empirical scope of the discipline is more productive than another in terms of making discoveries, developing useful theories, or advancing cultural values such as communal well-being. By making worldviews explicit researchers are better able to reflect on how their worldviews may impede, bias or facilitate both their personal scientific work and their collaborative efforts.

5.3.2   Data Questions

The primary knowledge in a discipline is the data collected in answer to Q4, namely "What are the subject entities like?" These answers represent pre-theoretical knowledge about the subject matter and is acquired via observation, experimentation and

---

[8] In many technical disciplines (for example Biology) the term "ontology" is used for what we have here called "subject terminology". In the context of a disciplinary Knowledge Base we will reserve "ontology" to refer to the set of entities treated as logical or natural objects by the discipline (more on this in Sect. 5.3.4.1), in line with conventional usage in philosophy. In the AKG model we also use the term "ontology" in the Guidance Framework in its philosophical sense, to refer to the worldview component concerned with what exists most fundamentally *in the universe* (relative to how the term "the universe" is construed in the worldview). The usage context should always make it clear in which sense the term "ontology" is being used.

pre-theoretical analysis. Disciplinary data represents the "morphology" of the subject entities, and it ranges over:

(i) descriptions that answer the question: *what are the features of subject entities?* These descriptions use terms from the subject terminology to characterize the properties of specimen subject entities in terms of observable characteristics such as form, structure, composition, functions, behaviours, powers, developmental stages, etc. These descriptions constitute what is known as "annotations".[9] Annotations index entities in ways that enable systematic classification (Mayor & Robinson, 2014, p. 688); and

(ii) an empirical classification that answers the question: *what are the relationships between the empirical subject entities?* The classification orders the knowledge about morphological characteristics in a theoretically neutral way, providing a "taxonomy". Taxonomies are typically based on (statistical) cluster analysis (Bailey, 1994), and yield categories called "taxons", representing what we will call "empirical kinds" (of subject entities).

5.3.3   Special Theory Questions

Q5 is concerned with the issue of "how do the subject entities work?". The availability of data makes possible the development of theories that explain or predict different aspects of the subject entities' morphology. This typically proceeds by iteratively addressing the question: *how is each kind of specific morphological feature produced or sustained?* The special theories identify the mechanisms involving processes that produce or sustain the morphological features (including functions) of the subject entities, and jointly constitute a construct we call "morphodynamics". In reflecting how things actually work, rather than merely how they appear to be or what they appear to do, special theories strive for increasing objectivity about the nature of their subject entities, and hence increasing independence from the perceptual constraints implicit in empirical studies. Finding out "how things work" is often thought of as the main or only objective of science, and developing such theories is indeed the major activity in most disciplines. However, scientific work extends also into a further and important area, namely the development of general theories. The development and structure of general theories is the main focus of the present chapter, and for this reason we will not here delve more deeply into how morphodynamic theories are developed.

---

[9] We adopt this technical term from its use in Biology to refer to associations between terminology ("ontology" in their usage) and biological entities (Hill, Smith, McAndrews-Hill, & Blake, 2008; Yon Rhee, Wood, Dolinski, & Draghici, 2008).

5.3.4 General Theory Questions

5.3.4.1 Q6: How Do the Empirical Entities Come About?

As will become clear, addressing Q6 results in work on the general theories of the discipline, because the answers involve factors relevant to the whole spectrum of subject entities. The reason for this is that in order to answer very general questions in a scientific way we have to transcend the issues around what the subject entities appear to be or what we believe them to be, and try to understand their intrinsic nature. It is by trying to understand entities from an objective or 'natural' perspective that we become increasingly able to match "nature's logic" in the building of explanatory theories about natural systems and systems built from naturalistic components.

It is in part because general theories deal with the intrinsic nature of subject entities that general theories provide common scientific grounds for the discipline, and hence scientific foundations for the unity of the discipline (unlike the empirical subject definition which, at least in young disciplines, provides more of an administrative or political unity).

5.3.4.2 Q6.1: How Do the Simplest Subject Entities Come About?

A pragmatic approach to answering Q6 is to begin by trying to explain how the simplest subject entities come about (Q6.1). We will indicate this subfield of morphogenetics by the term "protogenetics".[10] Q6.1 firstly drives us to first find out what the simplest subject entities are. In general, these have turned out to be *elemental* entities, i.e. all complex subject entities can be viewed as organized compositions of elemental entities. Examples of elemental entities are atoms in Chemistry and cells in Biology. In general, the elemental entities are also the prototypes of the subject entities, in the sense of being historically the first to exist.

The elemental entities are 'natural kinds', that is they represent natural or objective entities rather than empirical or subjective ones. The intrinsic nature of natural entities is determined by the nature of their parts and the nature of the processes that bring the parts into association (and, under some notions of emergence, by how the parts are organized). By such reasoning scientists conclude, for example, that chemical substances are (intrinsically) physical substances composed of physical atoms.

The elemental entities can be seen as concrete special instances of an abstract "general elemental model", for example the atom model in Chemistry and the cell model in Biology. For each kind of general elemental model there is typically a range of concrete instances, for example there are many types of atoms in Chemistry

---

[10]From the ancient Greek πρῶτος *protos*, meaning "first", and γενετικός, *genetikos*, meaning "genitive"/"generative", which in turn derives from γένεσις *genesis* meaning "origin". The noun "protogenetics" is not in current scientific use but the adjective "protogenetic" is in use in Geology to refer to mineral inclusions that are older than the host material.

and many types of cells in Biology. This range of types constitutes what we will call an "Elemental Ontology". A central task in formulating a theory about how the elemental entities come about is to work out the principles and mechanisms involved in bringing the parts into organized association. On the basis of having a general elemental model and principles for element production we can work out what types of elements could possibly exist, and the conditions for their synthesis.

In virtue of the elements sharing a common architecture there are natural relationships between their natural or intrinsic properties, and this can be demonstrated via a kind of classification known as a "Typology" (Bailey, 1994). "Types" are natural or logical categories of entities, in contrast to empirical or statistical kinds which are "taxons", as explained earlier in relation to taxonomies (Sect. 5.3.2). Typologies order knowledge in a scientifically significant way, revealing patterns of intrinsic properties across 'types'. Examples of typologies classifying elemental types are the Standard Model in Physics and the Periodic Table in Chemistry. These typologies can be important research instruments, suggesting avenues for scientific exploration and discovery, for example as we previously noted about the early Periodic Table.

The general elemental model forms an important 'building block' in the building of general theories about subject entities. It 'evolves' as the discipline matures, and progress with it can radically change the discipline's perception of the nature and potential of its subject entities. Typically the generic elemental model proceeds through a series of increasingly sophisticated models that at each stage have significant implications for the general theory of the discipline and for advances in the special theories. For example, in Chemistry the atom model progressed through the 'spherical impenetrable marbles' model starting with Democritus and still defended by Newton, to the Rutherford-Bohr model of a positively charged nucleus with electrons in surrounding circular orbits, to the Quantum Mechanical model in which the orbitals are modelled using the wave mechanics developed by Schrödinger. Each advance revolutionized our understanding of the nature of empirical chemical substances, and these in turn brought about dramatic advances in technologies and applications that employ chemical substances. Likewise, in Biology the discovery of DNA revolutionized our understanding of the nature of cells, the potentials of simple and complex organisms, and of the evolutionary relationships between organisms.

There is in this history an interesting implication for the systems sciences, where the notion 'system' is still far from settled. Typically, for any discipline the understanding of the nature of its subject entities is ultimately determined by advances in its general theory (especially in its Elemental Ontology), rather than by debates between groups with different perspectives or vested interests. We can therefore expect the issue of what 'system' represents to be resolved via a series of advances in general systems theories, and (given the lessons of history) we should expect to be both surprised and empowered by each of the advances.

5.3.4.3 Q6.2: How Do Complex Entities Come About?

Once we have an understanding of the elemental entities, we have scientific foundations for developing theories about how complex entities develop from simpler beginnings. The study of developmental mechanisms and processes is called "ontogeny" or "ontogenetics", literally meaning the study of the origin or mode of production of what exists.[11] We are here generalising this term from its use in Biology where it refers only to the origins of individual *organisms*.

Examples of ontogenetic theories are those of nucleosynthesis in Physics and embryology in Biology. Ontogenetic theories not only explain the origins and life cycles of individual entities but also enable us to distinguish between entities that are of different types and ones that are merely life-stage forms of a given type, and to explain what types of entities are possible.

Having identified the range of possible complex types we can explore the relationships between them using typological classifications, once again revealing patterns of properties across types that can be useful for developing either explanatory or exploratory research. Examples of complex entity typologies include Phylogenetic Trees in Biology and the Hubble Galaxy Sequence in Astronomy.

5.3.4.4 Q6.3: How Does the Variety of Complex Entities Come About?

Developmental theories explain some of the observed diversity of subject entities, by showing how some of this variety reflects merely different life cycle stages of the same type of entity. However this does not account for all the observed diversity, indicating the existence of multiple developmental trajectories with different kinds of starting points. This calls for theories explaining mechanisms whereby new types of entities can arise from existing types. Typically this would involve mechanisms that modulate existing morphodynamic and ontogenetic processes to produce starting points from which new kinds of complex individuals can be developed. An example is Darwin's theory of evolution by mutation followed by natural selection of the best adapted individuals. Evolutionary mechanisms and processes are studied under a construct called "phylogenetics",[12] and we propose that this term be applied in a wide sense referring to the origins of all kinds of "tribes" of types of entities, not just "tribes of organisms" as is the current practice.

---

[11] From the Greek ὄν, *on* (gen. Ὄντος, *ontos*), meaning "being; that which is", and from –γένεια, –*geneia*, meaning "mode of production" or γενετικός, *genetikos*, meaning "genitive"/"generative", which in turn derives from γένεσις *genesis* meaning "origin".

[12] From the Greek φυλή, φῦλον - *phylé*, *phylon* meaning "tribe, clan, or race" and from γενετικός, *genetikós* meaning "origin, source, or birth".

5.3.4.5   Q6.4: Why So Some Entity Types or Design Patterns Not Arise or Not Persist?

Developmental theories explain how complex individuals come about, and evolutionary theories how kinds of complex individuals come about. This leaves unanswered another general question about the subject entities, namely: *why, amongst all the variety of kinds and types that are theoretically possible, do some types and patterns not arise or not persist?*

Another way of posing this question is to ask: *why do things work as they do?* In general, there are many different organizational patterns that can perform the same functions, so why do the things that actually exist have the specific functional patterns[13] they do? Such issues are studied under a construct we will call "axionetics".[14] Axionetics studies the interactions between the productive, developmental and evolutionary possibilities of specific designs and the constraints entailed by operational or existential contexts. This balancing interplay can be seen both in the genesis of natural systems and engineered systems.

To answer Q6.4 we have to develop theories about how things are "channelled" towards the forms they attain by mechanisms or processes that seek to optimize designs not only for functional parameters but also against system "value criteria" such as efficiency, effectiveness, low resource reqyuirements, and ease of repair. Such criteria can only be met by taking into account the various kinds of environments in which the entity operates, and it is by being optimized for their complex environmental contexts that entities acquire traits such as resilience, robustness, evolvability and so on. Regarding natural systems there is a shift in perspective here from the orthodox evolutionary one, in which change originates in the individual (for example genetic mutation or intellectual creativity) and the environment then selects for survival: the axionetic perspective recognizes that the environment is also a complex source of change, that the entities have adaptive or regulatory powers that can facilitate or oppose the selection features in its environments, and that environments can be modified in significant ways by the dynamics of embedded entities. In this light we can see that entities and environments evolve together via an interplay of their respective change mechanisms and respective adaptive/regulatory mechanisms, and resilience comes from optimization over the design of the entity-environment whole. For engineered systems their design is likewise modulated by a trade-off between technical capabilities and socially-determined contextual value criteria.

Examples of axionetic studies include Epigenetics in Biology, debates about the Anthropic Principle in Cosmology, heuristics for Value-driven Systems Engineering,

---

[13] The term "functional pattern" is used here in a neutral way, and does not entail the existence of an intentional designer for whom the 'function' has some 'purpose'.

[14] Literally, "the art of producing value" from the Greek ἀξίᾱ, *axiā*, meaning "value, worth" and the related ancient Greek terms τέχνη, tékʰnɛ:, meaning "craftsmanship, craft or art", τεχνικός, tekhnikós, meaning "of or pertaining to art, artistic, skilful" and τίκτειν, tíktein, "to bring forth, produce, engender".

and the praxeology of Evolutionary Systems Design.[15] The models involved in axionetics can provide deep insights into why things have turned out as they have, and what might be viable possibilities for the future. In complex nature it is the axionetic principles that underpin viability, resilience and sustainability, and in complex socio-ecological scenarios it is through the understanding of axionetic principles, and the use of axionetic models, that we can have a reasonable hope of minimizing unintended consequences of our actions, interventions and technological productions.

Axionetic theories see subject entities in terms of wholes comprising subject entities embedded in environments in a way that drives continuous optimization of entity designs. This represents an important shift in the scientific perspective, in recognizing that entities cannot be properly understood when studied without consideration for the optimality of their designs for different systemic environments/contexts. It is of course the case that developmental and evolutionary theories also take environments into account, but the focus here is different: developmental theories tell us about the processes that build things that can perform specific functions, and evolutionary theories tell us about processes that enable the building of things that can perform new kinds of functions, but axionetic theories tell us how the designs of working things become elegant, and hence how entities become robust and resilient even though their environmental context can fluctuate over their lifecycle. Axionetic theories are explicitly systemic theories, whereas in general the systemic nature of scientific theories is not always so overt.

Axionetic theories lead us to recognize a special kind of systemic whole in nature, where entities are simultaneously parts in what can be modelled as multiple overlapping contexts, are optimized for balance with each type of context, and continuously rebalancing their optimality in the face of continuous change. This kind of whole represents a kind of nested panarchy,[16] rather than the hierarchical wholes of more classical approaches. There are multiple ways in which entities can be configured to achieve an elegant harmony with their multiple environments, and multiple ways in which a panarchy can adjust to balance the activities of dynamic embedded parts. Ecosystems are examples of such wholes, and an Ecosystem Panarchy provides an example of a typology of axionetic wholes.

The axionetic perspective on systems brings into focus an aspect of systemness that has received insufficient attention in terms of theory development.[17] Current debates about the nature of systems are often focused on systems more-or-less as they are at a moment in time, for example debates about emergence consider the consequences of the concurrent relationships amongst present parts of the system,

---

[15] For more on Evolutionary Systems Design, see (A. Laszlo, 1996b, 2001).

[16] The term "panarchy" in the sense used here refers to a form of organised complexity that involves a totality that has multiple interrelationships without forming a hierarchy, so that the totality evolves but not under some kind of linear dominance relationship. See (Gunderson & Holling, 2001).

[17] Valuable theoretical work in this area can be found in E. Laszlo (1987, 1994, 1996a) and A. Laszlo (1996b, 2001).

and Koestler's famous "holon" model (Koestler, 1967) considers a system's properties as determined by a balance between the consequences of the concurrent interactions between the parts of the entity and the concurrent interactions of the entity with its environment. These perspectives tend to de-emphasize the idea that the diachronic evolution of systemic designs towards increasingly elegant configurations embedded in (consequently) increasingly stable environments is an inherent rather than coincidental aspect of systemness. This is a regrettable neglect, for axionetic theories of systemness may provide us with the most profound insights into the nature and potentials of systems.

5.4   Overview of the Model of a Disciplinary Knowledge Base

The progression by which disciplinary knowledge is built up (at least up to general theories), as presented in the previous section, is summarized in the table given in Fig. 5.5. As Fig. 5.5 illustrates, the build-up can be viewed as progressing via a series of stages, each producing a distinct kind of disciplinary content. These are not independent activities but each builds on the previous ones, and all the previous ones are revisable in the light of a present one's findings. In the table there is only one row for special theories but ten rows for general theory components. This appears to bias the model in favour of general theories, but this is an illusion. The rows representing general theory components typically refer to one or few output instances, while the row representing special theories represents a potentially vast number of specialised theories (the special theories make up the bulk of any discipline's theories). We have here expanded the stages through which general theories develop because that is the prime focus of the present chapter. The earlier questions can easily be expanded into subsidiary ones but given present purposes we leave that task for another occasion.

Using this information we can also now also represent the cumulative build-up of a disciplinary Knowledge Base diagrammatically as given in Fig. 5.6. The format of Fig. 5.5 allows us to include more explanatory notes, while the format of Fig. 5.6 enables us to better illustrate the relationships between the components. Consistently with the preceding discussion we also show the guidance framework components that knowledge generation depends on (and informs in return).

5.5   Modelling the Structure of GST*

Following this model, we can now represent the data and theories of Systemology in the same way, as answering the same general questions *mutatis mutandis*, for example What counts as a system? What are systems like? How do systems work? How do systems originate, develop, and evolve? Why are some kinds of systems more resilient than others? This then generates the same knowledge components with the same relationships between them as in the generic model. This leads to a structure for Systemology's knowledge base as shown in Fig. 5.7.

There is a close correlation between the components and structure of Billingham's model presented earlier (Fig. 5.2) and the components of the 'general systems theories' element of the knowledge base as modelled in Fig. 5.7. The new model affirms the aptness of the distinction Billingham introduced between the generic model of an elemental system and the generic model of a complex system, her argument that there are spectrums (and hence typologies) of each kind, and her view that GST* is a much richer construct even than what is suggested by the idea of a "Periodic Table of Systems".

There are also interesting differences. Figure 5.7 presents a more generic model, in that the 'principles' components in Fig. 5.3 have been replaced with components representing theories that subsume principles and generic models of types of systems. There is also more detail in Fig. 5.7, having two components (rather than one) generating models for types of complex systems, and an extra 'layer' representing axionetic systems. Identifying this extra layer is an important addition, as a theory of axionetic systems may one day provide the most profound conception of the nature of natural systems in particular and resilient systems in general.

Figure 5.7 includes important preliminaries not explicitly shown in Fig. 5.2, namely the subject definition, subject terminology and worldview that developing a GST* depends on. The terminology aspect was however mentioned in both Billingham's and Boulding's papers. In *Skeleton* Boulding associated his system types with "levels of discourse", and in her previously cited works Billingham discussed the difficulties inherent in interpreting Boulding's model as due to inadequate development of the technical vocabulary available in Boulding's time.

In our view the new structure presented here, which further extends Billingham's extension of Boulding, provides us with a useful framework for developing GST*, as it frames the required elements in detail and in familiar terms, in a way that is consistent with the general structure of general theories in orthodox disciplines. The structure allows us to make a more detailed inventory of what we already have in hand, devise targeted strategies for developing the components in a systematic way, and make a more confident assessment of the potential value of GST*.

Two aspects of this new model stand out for us as particularly important.

First, we refined the important distinction, introduced by Billingham, between a spectrum of types of elemental systems and a spectrum of types of complex systems. In our view, this resolves one of the historical problems with interpreting Boulding's hierarchy. Billingham interprets Boulding's intent as suggesting a spectrum of elemental systems exhibiting increasingly complex behaviours, but many others have seen it as a hierarchy of increasingly complex systems. In the schema presented here we establish a principled place for both the elemental and the complex systems typology.

Second, we delineated a third spectrum of system types, namely axionetic wholes. It may be that this kind of general systems theory may one day offer the most profound understanding of systemness in nature, and hence reveal the most valuable principles, models and insights towards the design of resilient socio-ecological and socio-technological systems and the minimization of unintended consequences in systemic interventions and systems engineering.

5.6 Completing the Diagram of the Structure of a Disciplinary Knowledge Base

Earlier we modelled the key components of a Knowledge Base relevant to general theories and their precursors. However, as discussed in Sect. 5.2, a complete knowledge base also includes kinds of hybrid theories and kinds of methodologies. We will now briefly discuss these extra components and then update the model of a knowledge base, so as to provide a complete view of the structure of a disciplinary knowledge base. This is an easy extension of the model already provided, but as we will discuss at the end this provides a model with significant additional utility.

5.6.1 Hybrid Theories

The world is complex in ways that do not always fit into the neat categories of the individual disciplines, and this creates the need for interdisciplinary work. Such work eventually leads to hybrid theories representing a synthesis of two or more specialised perspectives on a single phenomenon, for example neuropsychiatry or biochemistry (for a more detailed discussion, see Chap. 3. Each one of the special and general theories of a discipline can become the basis for such an interdisciplinary synthesis, significantly extending the utility of a discipline.

5.6.2 Methodologies

In a discipline methodologies typically arise as soon as data collection starts, providing heuristics for action and technological applications (for example selective breeding programs appearing before scientific theories of evolution, and pottery glazes developed before scientific theories of Chemistry). Once theories are developed each kind of theory can provide insights which can facilitate the development of new/improved methodologies for new/improved kinds of interventions and technological applications. This holds for special, general and hybrid theories. As with the development of theories, methodologies develop cumulatively as data collection and theory development expands, and as with theory development 'earlier' methodologies can be significantly revised on the basis of results derived from the application of 'later' methodologies and advances in 'later' theoretical frameworks.

5.6.3 Complete Structure of a Disciplinary Knowledge Base

We can now add the hybrid theories and methodologies to Fig. 5.7 to produce the complete model as shown in Fig. 5.8.

With this structure in hand we can now draw the same diagram for the structure of the Knowledge Base of Systemology, as shown in Fig. 5.9.

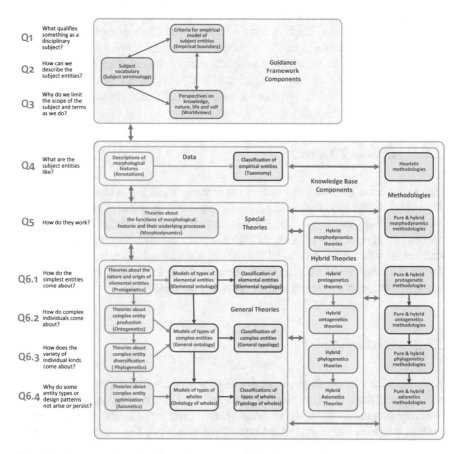

Fig. 5.8 The structure of a disciplinary Knowledge Base. (Reproduced from (Rousseau, Billingham, et al., 2016), with permission)

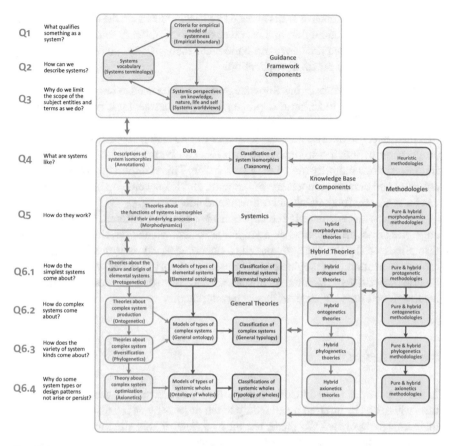

Fig. 5.9 The structure of Systemology's Knowledge Base. (Reproduced from (Rousseau, Billingham, et al., 2016), with permission)

5.7   Potential Value and Uses of the Detailed Knowledge Base Model

The model of a Disciplinary Knowledge Base developed here can be of value in multiple ways, especially to nascent disciplines and for the study of phenomena not yet scientifically understood.

Firstly, it can be used as a framework for making an inventory of current knowledge holdings. This can be useful in several ways, for example:

- putting the work of different researchers into context relative to each other, thus identifying connections that suggest opportunities for productive synthesis or collaboration;
- identifying key knowledge gaps in the discipline in support of formulating strategically prioritized research agendas; and

- defining the resources or skills needed to extend or leverage the disciplinary data, theories or methodologies, and hence to define disciplinary roles and associated skills matrixes required for data collection, theory development, methodology development and practical application.

Second, the structure of the Knowledge Base can be used to develop a classification framework for indexing disciplinary knowledge, making it accessible to systems researchers in a principled way.

Third, because the structure reflects work of increasing sophistication due to the progression of inquiry-driving questions, this can be used as the beginnings of a framework for a maturity model of the discipline. This is important for assessing the potential and current competence of the field, and for developing appropriate agendas for research, fundraising, recruitment and education.

Fourth, because it reveals the dependencies between different kinds of knowledge, the structure can serve as a guideline for knowledge development in young disciplines or for the study of puzzling phenomena, so as to avoid squandering resources on attempting sophisticated 'late stage' theory building when 'early stage' theories are still very immature or absent, and to ensure that appropriate foundations are developed for each stage of theory development.

5.8 Summary

In this chapter we argued that the general theory of a discipline has a generic structure, which can be inferred by attending to the process by which disciplines build up their knowledge base. We presented an overview of this structure, based on the sequence of questions a discipline addresses as it matures. We used this framework to envision the structure and scope of GST*. This provides a richer model of the scope and value of a general theory of systems than has existed previously. This model could provide useful guidance for the development of principles, models and theories of systemology.

In the next chapter the insights developed in this and previous chapters are brought together to develop ideas for discovering scientific general systems principles towards a GST*, and these are then applied to formulate an initial group of such principles.

References

Bailey, K. D. (1994). *Typologies and taxonomies: An introduction to classification techniques.* Thousand Oaks, CA: Sage.
Bard, J. B. L., & Rhee, S. Y. (2004). Ontologies in biology: Design, applications and future challenges. *Nature Reviews Genetics, 5*(3), 213–222.

Billingham, J. (2014a). *GST as a route to new systemics.* Presented at the 22nd European Meeting on Cybernetics and Systems Research (EMCSR 2014), 2014, Vienna, Austria. In EMCSR 2014: Civilisation at the crossroads – Response and responsibility of the systems sciences. Book of Abstracts (J. M. Wilby, S. Blachfellner, and W. Hofkirchner, Eds) (pp. 435–442). Vienna: EMCSR.

Billingham, J. (2014b). *In Search of GST.* Position paper for the 17th conversation of the International Federation for Systems Research on the subject of 'philosophical foundations for the modern systems movement', St. Magdalena, Linz, Austria, 27 April–2 may 2014, pp. 1–4.

Billingham, J. (2015). GST* as the unifying theory of the systems sciences. In D. Rousseau, J. Wilby, J. Billingham, S. Blachfellner (Eds.), *Systems philosophy and its relevance to systems engineering.* Workshop held on 11 July 2015 at the International symposium of the international council on systems engineering (INCOSE) in Seattle, Washington, USA. https://sites.google.com/site/syssciwg2015iw15/systems-science-workshop-at-is15

Bogdanov, A. A. (1913). *Tektologiya: Vseobschaya Organizatsionnaya Nauka* [Tektology: Universal organizational science] (3 vols). Saint Petersburg: Semyonov' Publisher.

Boulding, K. E. (1956). General systems theory – The skeleton of science. *Management Science, 2*(3), 197–208.

Bunge, M. (1979). *Ontology II: A world of systems.* Dordrecht: Reidel.

Bunge, M. (2014). Big questions come in bundles, hence they should be tackled systemically. *Systema, 2*(2), 4–13.

Francois, C. (2006). Transdisciplinary unified theory. *Systems Research and Behavioral Science, 23*(5), 617–624.

Friendshuh, L., & Troncale, L. R. (2012). *Identifying fundamental systems processes for a general theory of systems.* In Proceedings of the 56th annual conference, International Society for the Systems Sciences (ISSS), July 15–20 (23 pp.) San Jose State University

Gillett, C., & Loewer, B. (Eds.). (2001). *Physicalism and its discontents.* New York: Cambridge University Press.

Gunderson, L. H., & Holling, C. S. (Eds.). (2001). *Panarchy: Understanding transformations in human and natural systems.* Washington, DC: Island Press.

Hill, D. P., Smith, B., McAndrews-Hill, M. S., & Blake, J. A. (2008). Gene ontology annotations: What they mean and where they come from. *BMC Bioinformatics, 9*(5), 1–9.

Koestler, A. (1967). *The ghost in the machine.* Chicago: Henry Regnery.

Laszlo, E. (1987). *Evolution: The grand synthesis.* Boston: New Science Library.

Laszlo, E. (1994). An introduction to general evolution theory. *Journal of Biological Systems, 2*(1), 105–110.

Laszlo, E. (1996a). *Evolution: Foundations of a general theory.* Cresskill, NJ: Hampton Press.

Laszlo, A. (1996b). *Evolutionary systems design: Way beyond the two cultures.* In Proceedings of the Conversation on the comprehensive redesign of societal systems, International Systems Institute, Pacific Grove, CA.

Laszlo, A. (2001). The epistemological foundations of evolutionary systems design. *Systems Research and Behavioral Science, 18*(4), 307–321.

Martin, J. (2011). *INCOSE Systems science charter.* INCOSE. Retrieved from http://www.incose.org/docs/default-source/wgcharters/systems-science.pdf

Mayor, C., & Robinson, L. (2014). Ontological realism and classification: Structures and concepts in the gene ontology. *Journal of the Association for Information Science and Technology, 65*(4), 686–697.

Merriam-Webster. (n.d.). *Definition of MORPHOGENETIC. Merriam-Webster Online Dictionary.* Retrieved from https://www.merriam-webster.com/dictionary/morphogenetic

Mingers, J. (1997). Systems typologies in the light of autopoiesis: A reconceptualization of Boulding's hierarchy, and a typology of self-referential systems. *Systems Research and Behavioral Science, 14*(5), 303–313.

Mobus, G. E., & Kalton, M. C. (2014). *Principles of systems science* (2015th ed.). New York: Springer.

Rousseau, D. (2011). *Minds, souls and nature: A systems-philosophical analysis of the mind-body relationship in the light of near-death experiences* (PhD thesis). University of Wales Trinity Saint David, Lampeter, Wales, UK.

Rousseau, D., Billingham, J., Wilby, J. M., & Blachfellner, S. (2016). In search of general systems theory. *Systema, Special Issue - General Systems Transdisciplinarity, 4*(1), 76–92.

Rousseau, D., Wilby, J. M., Billingham, J., & Blachfellner, S. (2016). A typology for the systems field. *Systema, Special Issue - General Systems Transdisciplinarity, 4*(1), 15–47.

Simon, H. A. (1962). The architecture of complexity. *Proceedings of the American Philosophical Society, 106*(6), 467–482.

Troncale, L. R. (1978). Linkage propositions between fifty principal systems concepts. In G. J. Klir (Ed.), *Applied general systems research* (pp. 29–52). New York: Plenum Press.

Troncale, L. R. (1986). Knowing natural systems enables better design of man-made systems: The linkage proposition model. In R. Trappl (Ed.), *Power, autonomy, utopia* (pp. 43–80). New York: Plenum.

Troncale, L. R. (1988). The systems sciences: What are they? Are they one, or many? *European Journal of Operational Research, 37*(1), 8–33.

von Bertalanffy, L. (1932). *Allgemeine Theorie, Physikochemie, Aufbau und Entwicklung des Organismus (Theoretische Biologie— Band I)*. Berlin: Gebrüder Borntraeger.

von Bertalanffy, L. (1950). The theory of open systems in physics and biology. *Science, 111*(2872), 23–29.

Wilby, J. M. (2006). An essay on Kenneth E. Boulding's general systems theory: The skeleton of science. *Systems Research and Behavioral Science, 23*(5), 695–699.

Yon Rhee, S., Wood, V., Dolinski, K., & Draghici, S. (2008). Use and misuse of the gene ontology annotations. *Nature Reviews Genetics, 9*(7), 509–515.

Chapter 6
Scientific Principles for General Systemology

Abstract In this chapter, we develop a detailed discussion about the nature and evolution of general principles in science, how they relate to worldviews, laws, theories and theoretical virtues. We then apply this analysis to systems principles, to develop strategies for deriving general scientific systems principles. By applying these insights and also the conclusions from previous chapters we then present work done to discover three general scientific systems principles, and discuss some of the practical implications of these principles.

Keywords Heuristic systems principles · Scientific systems principles · The PLT model · The conservation of properties principle · The principle of universal interdependence · The principle of complexity dominance

6.1 Introduction: The Search for General Scientific Systems Principles

The foundations laid in Chaps. 1, 2, 3, 4 and 5 enable us to now start looking in a scientific way for the principles von Bertalanffy called for. He proposed that:

> It seems legitimate to ask for a theory, not of systems of a more or less special kind, but of universal principles applying to systems in general. In this way we come to postulate a new discipline, called General System Theory. Its subject matter is the formulation and derivation of those principles which are valid for "systems" in general (von Bertalanffy, 1956).

Progress in discovering such general principles has been slow, but the need for them has become increasingly urgent as the systems we would build, govern or nurture have become more complex. Without such principles the costs and risks of complex systems projects will increase to unsustainable levels (Augustine, 1987).

Major US defence systems projects typically overrun by about 50% (Lineberger, 2016) and big civil systems projects often overrun by 200% or more (Flyvbjerg, 2014). Two thirds of big IT projects fail, and more than half of those that are completed under-deliver on their promised value (Jones, 2016). The global cost of these failures and shortcomings is very large. In the USA, the cost of Systems Engineering

© David Rousseau 2018
D. Rousseau et al., *General Systemology*, Translational Systems Sciences 13,
https://doi.org/10.1007/978-981-10-0892-4_6

failures now exceeds $73 billion per annum ('Award#1645065 – EAGER/Collaborative Research: Lectures for Foundations in Systems Engineering', n.d.), and the global cost of IT project failures is now estimated at more $3 trillion per annum (Krigsman, 2012). Individual projects can fail even after very substantial investments: a recent US IT system project was abandoned after a spending of $100 million (Hardy-Vallee, 2012), and a recent UK IT system project was abandoned after a spending of £9.8 billion (Syal, 2013).

One response to these challenges has been renewed calls for advances in Systems Science, to more powerfully support the methods of Systems Engineering (SE) and Systems Practice. Such calls have recently been made in many stakeholder organizations, including the National Science Foundation (NSF), the International Council on Systems Engineering (INCOSE), the International Federation for Systems Research (IFSR), and the International Society for the Systems Sciences (ISSS) (Collopy & Mesmer, 2014; International Council on Systems Engineering (INCOSE), 2014; Rousseau, Wilby, Billingham, & Blachfellner, 2015; Wilby et al., 2015).

This call for advances in Systems Science has triggered renewed interest in systems principles and further calls for enriching the heuristic principles in current practice with more scientific ones. For example INCOSE, in their "Systems Engineering Vision 2025", said:

> It is therefore important to develop a scientific foundation that helps us to understand the whole rather than just the parts, that focuses on the relationships among the parts and the emergent properties of the whole. This reflects a shift in emphasis from reductionism to holism. Systems Science seeks to provide a common vocabulary (ontology), and general principles explaining the nature of complex systems (International Council on Systems Engineering (INCOSE), 2014).

These calls have stimulated recent debate about the nature, role and developmental status of systems principles, and this chapter is a contribution to that discussion. In particular, we will here argue that these questions should be addressed in the light of how principles are understood, used and discovered in academia generally, rather than exclusively building on the ideas of the founders of the systems traditions.

6.2   The Nature and Role of Scientific Principles

6.2.1   What Are 'Principles'?

A 'principle' is a fundamental idea or rule that can provide guidance for making a judgement or taking action. Principles can take the form of injunctions, beliefs, concepts, assumptions or insights. Principles can range from fully heuristic ones (distilled from experience, intuition, belief or convention) to fully scientific ones (distilled from scientific theories or models). Principles are encountered in every sphere of human activity, so we have for example principles relevant to ethics, aesthetics, economics, politics, science, engineering, agriculture, etc.

Examples of principles include the heuristic principle "do as you would be done by" and the scientific principle that "energy is conserved in all causal interactions".

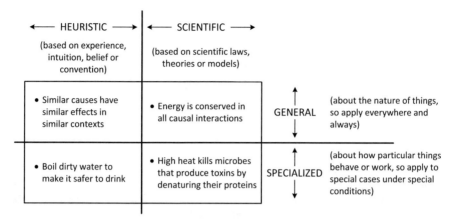

Fig. 6.1 Relationships between forms of principles. (Reproduced from Rousseau, 2018b, with permission)

Historically, principles start out as heuristics, and over time some become more scientific, for example Lucretius's heuristic principle from 75 BCE that "nothing can come from nothing (or go to nothing)" is today the scientific principle that "energy cannot be created or destroyed but only transferred or transformed". As principles become more scientific, they become more useful for making apt judgements or taking effective action.

By "more scientific" principles we mean principles that more strongly reflect the scientific approach, that is, use clear and precise concepts, express qualities and relationships that can be subject to measurement, quantification, empirical verification or falsification, and so on. In this sense scientific principles can arise in philosophy, science, engineering and operational/service contexts. The scientific enterprise can be viewed as aimed at making principles across these domains increasingly scientific. All domains that seek to develop or employ such principles can be considered to be scientific disciplines, becoming more scientific over time as their principles become more so.[1]

Note that we make a distinction between "scientific principles" in the sense just explained and "science principles", i.e. the principles underpinning science. It is a separate question whether the principles underpinning disciplines such as sociology, anthropology, economics, politics or psychology are scientific or not.

Both heuristic and scientific principles can be either general (applying universally, for example conservation of energy) or specialised (applying only in specific contexts, for example the principles of disease prevention), as illustrated in a simplified way in Fig. 6.1 with example principles.

The effectiveness of science depends on having strong principles underpinning scientific research methods, and the progress of science at a fundamental level (such

---

[1] In this sense many disciplines beyond the traditional 'hard sciences' can be considered to be scientific in spirit and becoming increasingly scientific in practice.

Fig. 6.2 The levels hierarchy and its emergence over time. (Reproduced from Rousseau, 2018b, with permission)

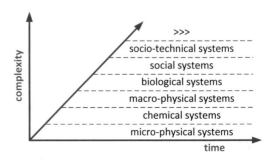

as the discovery of new substances or new laws of nature) depends on having strong general principles. For example, specialized laws of nature, for example Boyle's Law that states the balancing relationship between pressure and volume in an ideal gas, are instances of general principles such as that energy is always conserved or that effects have sufficient causes. General principles are powerful guides for exploring phenomena for which adequate theories do not yet exist.

6.2.2   What Are Systems Principles?

From the understanding of the nature of principles just presented we can now say that *systems* principles are fundamental rules, beliefs, ideas or insights about the nature or workings of systems, and hence systems principles guide judgment and action in systemic contexts. Systems principles will therefore exist in both heuristic and scientific forms, and in both general and specialised forms. Moreover, general scientific systems principles will have the same relevance for systems laws, and for exploratory systems research, as the relationship just described for the sciences more broadly.

The principles von Bertalanffy called for are *general* scientific systems principles (GSSPs), and to form a clear idea of what these might look like and how they would differ from specialised systems principles it is useful first to consider the nature and scope of System Science, with a view to understanding how principles both underpin and flow from Systems Science.

6.2.2.1   Systems Science and Its Relationship to Principles

A starting point for thinking about Systems Science is the view that every concrete thing is a system or part of one, and that natural systems can be arranged into a "complexity hierarchy", in which every level corresponds to some kind of system and the 'levels' represent increasingly complex systems embedding systems from the 'lower' levels, as shown in a simplified way in Fig. 6.2. A version of this perspective already occurs in Aristotle, but there is an extensive modern literature on this, for example (Poli, 2001; Rueger & McGivern, 2010) and notably, in the specific context of general systems theory, a seminal paper by Kenneth Boulding (, 1956).

The system levels in the complexity hierarchy correspond to the subjects of concern of the mainstream specialised scientific disciplines, so it can be said that every specialised scientific discipline studies some kind of system. Note however that this does not make these disciplines systems sciences, since it is only trivially true that their subjects are systems. These specialised disciplines do not have as their subject matter systems *as systems* but rather they seek to understand instances of kinds of systems.

The idea of a science of systems arises from three reflections on the complexity hierarchy:

1. First, given that systems occur on every level of the complexity hierarchy, a science of systems must be about what is true of or possible for systems across all the levels. This is the insight behind the claim that System Science will be a transdiscipline, having relevance across the disciplinary spectrum, and will comprise theories that are scale-free and composition-independent. At a minimum, such a science must involve concepts and principles that allow systems to be characterised as a category of analysis distinct from things that are *not* systems, to enable instances of systems to be identified in the real world, and to explain/predict the behaviour and potential of systems as *systems*. Our present notions of 'systemhood' are far from settled, but there is a rich literature on the subject (Ackoff, 1971; Bunge, 1979, 2003; Schindel, 2011; von Bertalanffy, 1950) (see also footnote 3) and important efforts are under way to consolidate these ideas (Sillitto et al., 2017, 2018).

2. Second, when looking across the levels we find similar patterns recurring across multiple levels, for example spiral forms in certain tropical storms, sea shells, flowers and galaxies. Other examples include Fibonacci sequences and Zipf's Law regularities in natural phenomena (Friendshuh & Troncale, 2012; McNamara & Troncale, 2012; Troncale, 1978). Speaking metaphorically, these patterns represent solutions to design problems that nature must solve in order to create enduring complex structures. The existence of these isomorphically recurring patterns across changes in scale and composition entails that there must be transdisciplinary specialised systems principles reflecting the nature of these 'solutions'. In principle each of these patterns can be 'decoded' to establish a theory that explains the nature and function of the observed pattern, and to identify the relevant explanatory principles. Each such theory would then be a specialised systems science theory, and we have several of these already (for example Control Theory, Hierarchy Theory, Network Theory, Communication Systems Theory, Theory of Dissipative Structures etc.). There are still many patterns in nature we do not theoretically understand, for example patterns of overlapping Fibonacci spirals, and Zipf's Law patterns. Moreover it is likely that there are further patterns we have not yet identified.

3. Third, the isomorphically recurring patterns arise independently in multiple contexts involving different scales, compositions and developmental histories. This suggests that there are general systems principles that provide for the possibility of the *emergence* of these systemic patterns across contexts. Speaking loosely, these would be general principles about how Nature 'finds' solutions, rather than (as above) specialised principles about how specific kinds of solutions work. We have very limited knowledge of such general systems principles,[2] but in principle they hold the promise of a general theory of systems that would explain both the emergence of specialized patterns and the relationships between them. Such a 'general systems theory' (GST*) would be very valuable not only for unifying the body of specialised systems knowledge but also for opening up new routes to discovery, just as Mendeleev's periodic table of elements did for Chemistry and Darwin's theory of natural selection did for Biology.

From this we can infer that the theoretical aspect of Systems Science minimally comprises a set of concepts used to characterise the universal attributes of systems as systems, a database of isomorphic systems patterns,[3] specialised systems theories that explain the mechanisms underpinning specific isomorphic systems patterns, and a general theory of systems that explains how the universal system attributes arise in nature and how they support the emergence of the isomorphic system patterns. *The insights entailed by these concepts and explanations are the general and specialised principles of Systems Science.* In practice Systems Science also includes the hybrid theories where systems principles are used in (or derived from) the study or modelling of specialised kinds of systems, for example Systems Biology, Systems Ecology, Systems Psychology, Systems Economics and so on. From these theories an additional range of *specialised* principles can be distilled.

We can now paraphrase Fig. 6.1 for the case of *systems* principles, as illustrated in Fig. 6.3. If we look at the principles available in Systemology at present, we can see that we have a large number of heuristic principles in hand, deriving from the many established methodologies within systems thinking and practice. For lists of such principles see for example (Hitchins, 1992, pp. 60–71; Mobus & Kalton, 2014, pp. 17–30; Walden, Roedler, Forsberg, Hamelin, & Shortell, 2015, pp. 20–21). It should be noted that systemists have also published many heuristic statements under the rubric of 'systems principles' or 'systems laws' without these statements being actually useful for making judgements or taking action. These heuristics are

---

[2] Early work on general systems principles focused largely on concepts (for example, (von Bertalanffy, 1969, pp. 91, 95)), and while these remain controversial, important progress is now being made (for example, (Sillitto et al., 2017)). In addition, progress is now being made towards establishing propositional general scientific systems principles. Two recent papers respectively presented three such principles (Rousseau, 2017b) and eight strategies for discovery projects (Rousseau, 2017a).

[3] Len Troncale and colleagues have over 40 years made an important contribution to the development of such a database of systemic isomorphisms, and extended this by also analysing the linkages between isomorphisms (Friendshuh & Troncale, 2012; McNamara & Troncale, 2012; Troncale, 1978, 1985, 1988).

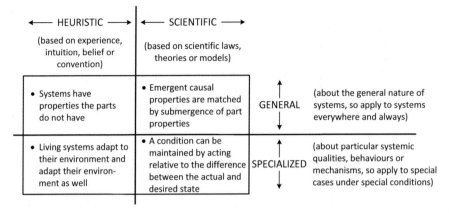

Fig. 6.3 Relationships between forms of systems principles. (Reproduced from Rousseau, 2018b, with permission)

typically just witticisms or platitudes about systems, such as "today's problems come from yesterday's solutions" (Senge), or "complex systems exhibit unexpected behaviour" (Gall). For lists of such principles see, for example, (Augustine, 1987; Gall, 1978; Senge, 1990). Others have published heuristic principles that are useful but not general, notably (Maier & Rechtin, 2009), which lists principles for specific contexts such as architecting, design, social systems, and political processes. For summaries of other heuristic specialised systems principles, see also (Katina, 2016; Pratt & Cook, 2018). We also have a small but significant collection of specialized scientific systems principles deriving from the dozen or so specialised systems theories such as control theory, hierarchy theory, fractal theory and so on. These are important for designing systems that are robust, reliable, efficient, effective and resilient.

When it comes to general scientific systems principles there are a small handful of general concepts (for example wholeness, part, equifinality, closed and open system (von Bertalanffy, 1969, pp. 91, 95).[4] These concepts are still far from settled, including even the concept of 'system' (Sillitto et al., 2017), and prior to the GSTD project no general scientific systems principles framed as propositional principles. Three such general scientific systems principles have since been identified based on the conceptual foundations laid by the GSTD project (Rousseau, 2017b), and their basis and scope will be explained later on. They however have yet to be formalized,

---

[4]There are very many concepts relevant to systems in the vocabulary of Systemology, for example there are 3807 entries in the second edition of Charles Francois' *International Encyclopedia of Systems and Cybernetics* (Francois, 2004). These terms are far from standardised, and many systemologists have produced their own lists, for example (Danielle-Allegro & Smith, 2016; Heylighen, 2012, pp. 21–33; Kramer & De Smit, 1977, pp. 11–46; Schindel, 2013; Schoderbek, Schoderbek, & Kefalas, 1990, pp. 13–68; Wimsatt, 2007, pp. 353–360). However, very few of these concepts are *general* systems concepts, i.e. concepts describing universal attributes of systems *as systems*.

but a project to do this and initial projects to evaluate them are in process (Calvo-Amodio, Kittelman, & Wang, 2017).

This pattern of available principles of course reflects the experiential richness but theoretical immaturity of our knowledge of systems.

To set us on a course to discovering general scientific systems principles, it is useful to first take a closer look at how general scientific principles feature in science more broadly, and then to transpose these insights to the context of systems.

6.2.3   The Role of General Principles in a Scientific Discipline (PLT Model)

There are multiple terminologies and perspectives in science and in philosophy on the nature of the relationships between general principles, laws, theories and models. For present purposes, we will follow a perspective called Scientific Realism, which is presently the dominant view amongst metaphysicians of science (Schrenk, 2016, p. 299), is well matched to the working practice of practicing scientists and is consistent with the General Systems Worldview as discussed earlier. Briefly, Scientific Realism posits that a concrete world exists independently of our mental states, that the truth of our theories depends on the nature of the world, and that our best scientific theories are approximately true of the world.[5] Within the framework of Scientific Realism we propose following a model known as the "Principles-Laws-Theories" (PLT) model of modern science (Dilworth, 1996). For present purposes we will focus only on its notion of principles. The PLT model represents an early attempt (1996) in the modern resurgence of metaphysics to show how modern science depends on metaphysical principles and how such principles relate to scientific laws and scientific theories. The metaphysics of science has advanced rapidly in the last two decades, but in our view the basic structure of the PLT model is still the most practically useful framing we have of these relationships. We will however expand the model to make the dynamics of scientific developments and the connection to worldviews more explicit.

In science, general principles articulate the most fundamental assumptions we make about the nature of the world. They represent what we take to be true in general, and hence fulfil a number of orienting functions, including (Dilworth, 1996, pp. 65–72):

(a) Encapsulating what is deemed ontologically or metaphysically possible or inevitable (for example, the "Principle of Sufficient Reason", which claims that

---

[5] Useful discussions of Scientific Realism can be found in (Chakravartty, 2010; Ellis, 2009; Schrenk, 2016). See also endnote 1. Within the present system movement it is related to the view called Critical Realism (Mingers, 2006, 2014). As a worldview component it is close to the view of the founders of the general systems movement, known as the General Systems Worldview, as discussed in Chap. 4.

effects have proportionate causes, is a presumption against the occurrence of miracles);

(b) Setting bounds of scientific forms of reasoning (for example the "Principle of Uniformity of Nature", which claims that under the same conditions the same causes always produce the same effects, presents one way in which evidence can be linked to conclusions or predictions);

(c) Providing guidelines for doing science (for example the "Energy Conservation Principle" provides a way of checking that the all the contributors to a given effect have been identified); and

(d) Defining basic concepts (for example, 'energy', 'force' and 'atom').

The principles are not independent claims but can overlap or reinforce each other. For example, the Principles of Sufficient Reason can be viewed as a corollary of the Principle of the Uniformity of Nature (or vice versa).

The principles of science are grounding *assumptions* and hence not provable by science. However, they are provisional and can be challenged and amended. Nevertheless, they are regarded as representing deep truths about the nature of the world, and their formulation and evolution is informed by progress in science. They express what we take to be the conditions for the possibility of the empirical phenomena observed by sentient beings. In this way the principles of science represent the invisible reality underlying the phenomenal one, and form part of metaphysics rather than science. Taken together, the principles of science characterize the nature of Nature, so we might say that our image of the nature of Nature is the gestalt that reconciles the joint entailments of the principles (rather like the elephant image that reconciles the observations of the seven blind men). These relationships are illustrated in a simplified way in Fig. 6.4. Changes in the principles can have dramatic consequences for the scientific paradigm, as for example occurred when the Newtonian notion of "mass" was redefined by Einstein's General Relativity theory.

Principles generally start out as qualitative heuristic principles based on limited observations, and later on (typically with great difficulty) become exact, quantifiable and profound. For example, the (heuristic) Aristotelian notion of a force defined a force simply as a push or a pull, while the (scientific) notion from Newton was quantitative and carried profound implications, triggering the "Mechanical Revolution".

Note that principles as stated here typically do not explicitly take the form of a guideline, framed as for example "if you want x in context y then do z". However, if the principle is understood it can be interpreted as a guideline and hence be applied in taking action or making a judgement. For example, the principle that energy is conserved in all causal interactions is equivalent to a guideline that says if we wish to scientifically explain an effect we can identify the causes by tracing the flows of energy along space-time tracks towards the final state. The same can be said for those *concepts* that are treated as principles. For example, the concept 'force' as defined by Newton is equivalent to a guideline that says if you want to

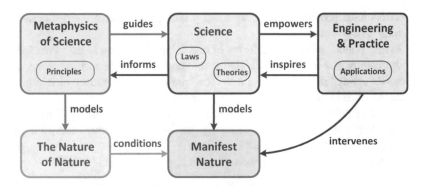

Fig. 6.4 The significance of philosophy for science and practice/engineering. (Reproduced from (Rousseau, 2017a), with permission)

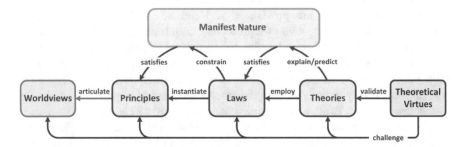

Fig. 6.5 The "Principles-Laws-Theories (PLT) model" of modern science. (Reproduced from (Rousseau, 2017a), with permission)

explain something's state of motion you have identify the balance of forces acting on it.

6.2.4   The Interdependence of Principles, Laws and Theories

Principles, laws and theories interdepend systemically, and this conditions how they are discovered, used and evolve. The "PLT model" mentioned earlier (Dilworth, 1996) captures these relationships well, as illustrated in Fig. 6.5 and explained below.

The guiding principles for doing science (for example that similar causes produce similar effects) express general assumptions or accepted general insights about the nature of the world, and therefore the general principles jointly form the most succinct expression we have of our worldview. Conversely, if we can describe our worldview we can distil general principles from it. Once we can state the principles we can apply them to observations of causal interactions to discover laws of nature, which are exemplars of the principles in specific contexts. For example, Boyle's

Law specifies how an increase in the pressure of an ideal gas will cause it to proportionately expand in volume, in conformance with the general principle that all effects have proportionate causes (under given conditions). Conversely, laws can be generalised to suggest new principles, for example Kepler's second law, which states that planetary orbits sweep out equal areas in equal time, can be generalised to suggest the Principle of the Conservation of Angular Momentum. By applying laws we have derived in this way to observations of previously poorly understood phenomena we can develop models and theories that explain or predict those phenomena. For example, we can apply Newton's laws of motion of massy objects to data from astronomical observations to build a theory that explains why we have two ocean tides per day, or to build a model that predicts the return of specific comets.

An interesting nuance is added by the fact that in practice there are often multiple ways of explaining the same phenomenon. To choose between them, competing theories or models are judged as to how "good" they are by evaluating them against "theoretical virtues" such as explanatory power, predictive power, simplicity, falsifiability, coherency, empirical adequacy, consistency with well-established theories (Matthewson & Weisberg, 2009; Maxwell, 2004). Philosophy of science has shown that theories that are 'good' in this sense are 'better' because they tend to last longer before they are superseded, are more likely to lead to new insights, are more likely to evolve into even more powerful theories rather than just be discarded, and so on.

If we cannot develop "good" theories about a given phenomenon, we must question the adequacy of the laws they employ: perhaps these need additions or refinements, or we need extra ones. To discover new or improved laws we have to reflect on our principles, because laws are special cases of how the principles play out under specific conditions. By making further careful observations of the puzzling phenomena, and then carefully applying our principles, we might find better or further laws, which we can then use to develop more powerful theories and models. If despite these efforts we still cannot devise 'good' theories, we must then cast doubt on our principles. We generally refine or extend them by generalising from laws we already have, or by distilling them from the assumptions entailed by our worldviews, so if we are questioning our principles we have to consider both possibilities. As a first step we could review our current stock of laws and reflect on whether there are opportunities for hitherto unforeseen generalizations that could help us build the better theory we need. In general, discovery often works this way, for example Kepler's second law was discovered before the generalized principle it instantiates (conservation of angular momentum) was known. However, if new or improved principles cannot be found in this way, or what we do find does not help us to improve/extend our laws such that we can build good theories, then we must question our worldviews, reflecting on how we balance between knowledge, experience and intuitions to find the core beliefs that ground our basic judgements and actions, and hence we must try to form an adjusted worldview from which we can then adjust or extend our principles, laws, theories and models.

In this manner, assessments against the theoretical virtue criteria help to drive the evolution of theories, laws, general principles and worldviews in a systemic way. If

Systems Science is a scientific endeavour then it will follow the same pattern of discovery and evolution as we search for scientific systems principles, laws and theories, so it is helpful for systemists to understand these interrelationships in our quest to find routes to discovering them.

As an aside, it may be useful to note that in order to effectively leverage this insight it is advisable to adopt the methods and language already in use in science and philosophy to model this process and capture its outcomes, so that we can maximise the lessons we can learn from established science and the metaphysics of science, and minimise the effort needed to integrate the findings of Systems Science back into the established body of science. For example, accepting "scientific principles" as denoting our most general assumptions about the nature of the world, then entails that scientific "systems principles" express our most general assumptions about the systemic nature of the world.

In what follows, we will apply these ideas from science and the PLT model to foundational systems concepts, in order to derive insights for foundational discoveries in systems science. As a first step, we will paraphrase the analysis given above for the systems scenario.

6.2.5   The Nature and Significance of General Systems Principles

The content of Systems Science is distinct from that of the specialized sciences, but the structure of Systems science is likely to be no different from that of the rest of science. From this brief review we can thus form some idea of the scope and potential of systems principles. We can directly paraphrase the above discussion for the systems case as illustrated in Fig. 6.6.

The correspondence between these two diagrams lie in the observation that Systems Philosophy models the systemic nature of the nature of Nature, and Systems Science models the systemic nature of manifest systems.

Paraphrasing what was said above about our image of Nature, we can now propose the following. General systems principles are the grounding assumptions of systems science, and hence not provable by systems science. However, they are provisional and can be challenged and amended. Nevertheless, they are regarded as representing deep truths about the systemic nature of the world, and their formulation and evolution is informed by progress in systems science. They express what we take to be the conditions for the possibility of the empirical systemic phenomena observed by systems thinkers. In this way the systems principles represent the systemic nature of the invisible reality underlying the systemicity of the phenomenal one, and form part of systems philosophy rather than systems science. Taken together, the systems principles characterize the nature of systemness, so we might say that our image of the nature of 'system' is the gestalt that reconciles the joint entailments of the systems principles. A set of coherent and scientific systems prin-

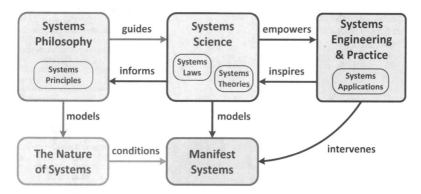

Fig. 6.6 The significance of Systems Philosophy for systems science and systems practice/engineering. (Reproduced from (Rousseau, 2017a), with permission)

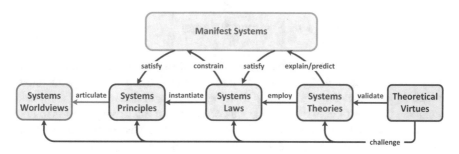

Fig. 6.7 A "Principles-Laws-Theories (PLT) model" of systems science. (Reproduced from (Rousseau, 2017a), with permission)

ciples would form the core of a foundational general systems theory (GST*), and changes in the systems principles could have dramatic consequences for the systems worldview.

Once we have some principles in place for a scientific GST*, we would be able to execute a cycle of discovery, progress and refinement in the context of systems science, in the same pattern as discussed above for the PLT model more general. We illustrate this in Fig. 6.7.

6.3 Foundations for Discovering General Scientific Systems Principles

6.3.1 Grounding Concepts for a Search for General Scientific Systems Principles

In what follows, we will show how these broader ideas can be applied to the systems science context, and lead to useful discoveries.

Fig. 6.8 A hierarchy of kinds
of systems. (Adapted from
(Rousseau, 2011))

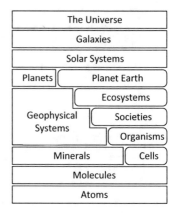

6.3.2   General Ideas in Systems Science

A common idea in systems thinking, as discussed earlier in relation to Fig. 6.2 is
that we can arrange naturalistic systems into a hierarchy by sorting things into kinds
based on properties that are essential to being members of that kind, and then rank-
ing them in order of complexity. One way of looking at this arrangement is to note
that the things in every layer are composed of things that exist autonomously at the
'lower' level. Figure 6.8 gives another impression of this.

The "layers" are usually referred to as "levels of reality", although it may be
more apt to call them 'levels of description'. Seeing it as a hierarchy in this way
corresponds to the reductionistic idea that systems on one level have as their parts
systems from the lower levels, and that properties of the systems at any level can be
explained in terms of the properties of, and the relationships between, their parts.
There are obvious controversies around these considerations, but for present purposes we are only concerned with highlighting some basic system concepts involved
in constructing such a model.

Systems differ from heaps in that the properties of heaps are merely the sum of
the properties of the parts, whereas systems have new kinds of properties their parts
do not have. These are called "emergent properties", and it is this emergence of new
kinds of properties that establishes new kinds of systems (Bunge, 2003). One definition of "system", due to Anatol Rapoport, is that a system is a whole that functions
as a whole in virtue of the relationships between its parts (Rapoport, 1968). When
the systems at stake are naturalistic ones, as shown in the complexity hierarchy, then
the inter-part relationships that establish the whole must be concrete, and therefore

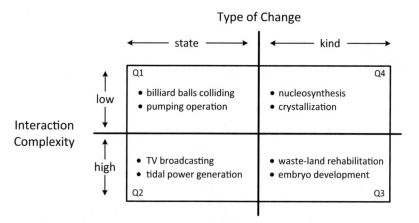

Fig. 6.9 Four categories of systemic interactions (with examples). (Adapted from Rousseau, 2018a)

must be due to lawful causal interactions between the parts.[6] In this light we can say that emergent properties are new kinds of causal powers that arise at the level of the whole due to kinds of causal interactions between parts. Systems can have a multiplicity of kinds of parts, and a multiplicity of inter-part relationships, leading to a multiplicity of kinds of interactions between the parts. As the diversity of parts, relationships and interactions increases systems are said to become more "complex" (Schöttl & Lindemann, 2015). That said, it is important to note that interactions between parts do not always produce new kinds of systems – in fact for the most part interactions just create new states in existing systems.

We can model these intra-systemic interactions and their consequences in terms of two dichotomies. On the one hand, we can characterise intra-systemic interactions in terms of their complexity, by considering whether the diversity involved in the interactions is low or high. On the other hand we can characterise the interactions in terms of the kind of change they produce, specifically whether they produce state changes in a given system or produce a new kind of system through property emergence. We can illustrate these distinctions as indicated in Fig. 6.9, with some examples given.

On the left side of Fig. 6.9 (Q1 and Q2) we have interactions that merely produce changes of system state, either via mechanical interactions within simple systems (for example collisions between balls on a billiards table as shown in Q1) or a diversity of electro-mechanical processes within complex systems (for example the processes in a tide-driven electricity generator as mentioned in Q2). Although these are systemic interactions their workings are well understood within the existing science paradigm. On the right hand side of Fig. 6.9 we deal with interactions that produce emergent properties and hence produce new systemic identities, for exam-

---

[6]Things are "naturalistic" if they can only change in accordance with the laws of nature, and "concrete" if they have causal powers.

ple when a small number of fundamental physical particles combine to form a new atomic nucleus (Q4), or when a large variety of systems are brought together to establish a forest ecosystem where once there was waste-land (Q3). Interactions that produce new kinds of systems are largely mysterious in the existing science paradigm, and in this area the systems thinking approach could make an important contribution (see for example (Arnold & Wade, 2015)).

It is important to note that one of the challenges in trying to explain emergence comes from the fact that science's approach is predominantly reductionistic, taking new kinds of things to be largely explicable in terms of special states of collections of lower-level entities and so on 'all the way down' to fundamental particles (quantons). This was famously expressed by Steven Weinberg saying "all the explanatory arrows points downwards" and remarking that this is "perhaps the greatest scientific discovery of all" (Weinberg, 2008). Systems thinkers are typically sceptical about such a narrow reductionism, because living systems exhibit properties that are categorically different from physical ones, such as subjectivity and anticipation, and context can powerfully influence developmental processes, as seen in cultural inheritance. However, it has proved an enduring challenge to formulate a scientifically profound systemic alternative to reductionism, despite a vigorous philosophical debate on this subject (for example (De Caro & Macarthur, 2010; Gillett & Loewer, 2001; Koons & Bealer, 2010)).

We will now take these general ideas from the foundations of systems thinking and reflect on them in the light of the general principles foundational to science. We will explore how to express the former in terms of the latter, working from simple cases to complex ones, and see if we can discover insights about systems that are general, scientific and useful.

6.3.3 A First Strategy for Discovering GSSPs

The scientific principles that jointly characterize the nature of nature (as shown in Fig. 6.4), are expressed in terms of the foundational concepts in the metaphysics of science, and the desired scientific general systems principles (SGSPs) that would characterize the systemic nature of nature (as shown in Fig. 6.6) would be expressed in terms of the foundational concepts of systems science. This suggests that we might discover SGSPs by investigating how the core concepts involved in the principles underpinning science might be related to the foundational concepts of systems science, and then extrapolating SGSPs by applying the implications of scientific principles to relevant systems concepts.

6.4 Derivation of Three General Scientific Systems Principles

6.4.1 Emergence and the Conservation of Energy

As a first investigation, let us begin by considering the notions of properties and interactions firstly from a metaphysics and then from a systems perspective. On the metaphysics of science side it can be seen that scientific principles, and the laws that instantiate them, model changes in substances in terms of the sources and consequences of change, and the proportionalities between changes in different substances in different contexts. These models depend on the notions of causal powers, concrete properties, and exchanges or transfers of energy, using ideas we can briefly summarize as follows[7]:

1. 'Real' (non-imaginary) things are called 'concrete' if they have causal powers, and causal powers are properties that make interactions between things turn out in a specific way;
2. In interactions, changes occur in the interaction partners, so causal powers can be understood as the power to cause or undergo change;
3. If the changes are proportionate (i.e., effects are proportional to their causes) then the interactions are 'naturalistic';[8]
4. The ability to cause change is the ability to do work, and the doing of work requires transfer or transformation of energy;
5. This implies that to have concrete causal powers is to have energy and the ability to exchange it; and
6. In this way, there is a clear and direct connection between the notions of concrete properties, causal powers, change, and energy.

Under Scientific Realism the concept of energy has a precise meaning (Bunge, 2000), and empirically energy can be quantified in an exact way. This opens up an opportunity to make the notions of causally effective properties and empirical change exact and quantifiable too, as follows. If 'having causal powers' can be represented by 'having energy', then:

1. The kinds of causal powers something has can be represented by the kinds of energy it has;
2. The strength of a kind of causal power something has can be represented by the amount of the relevant kind of energy it has;
3. Causal interactions transfer and/or transform kinds of energies, and hence causal interactions can be viewed as changing the strengths of the interacting things' causal powers, and thus their concrete properties; and

---

[7]At least they can be from the perspective of Scientific Realism, the currently dominant view amongst metaphysicians of science (Schrenk, 2016, p. 299). For more on Scientific Realism, see (Chakravartty, 2013; Ellis, 2009).

[8]Note that these proportionalities do not have to be linear.

4. If causal interactions are naturalistic then energy is conserved during interactions.[9]

If we now turn to naturalistic systems, we might be able to speak of the emergent properties arising though interactions between parts in these terms involving causal powers and energy. To see how this might go, let us first consider a simple case, i.e. where the complexity of the interactions is low but they nevertheless produce an emergent property. This would be a case falling into Q4 of Fig. 6.9.

A convenient example is provided by atom formation, the process whereby protons and neutrons combine to form an atomic nucleus, as happens in stars and supernovae, and then combine with electrons to form atoms. The atom has properties the parts did not have – it is a stable structure that has different causal powers to those of the fundamental particles, and different quantities and arrangements of such parts will result in different levels of atomic stability. This stability is a new system-level property that emerges as the atom is formed. It is a concrete property, making a difference in causal interactions, and therefore is a causal power.

However, as explained earlier causal powers can be represented in terms of kinds of energy, so this entails that atoms have special kinds of energies that unbound fundamental particles do not. Now, we know from the principle of conservation of energy that 'emergent' energy must have come from somewhere. Given that the emergent property exists due to the interaction of the parts, it seems likely that the parts have given up some of their energy and hence have undergone a reduction in their own properties and causal powers. This is not an unreasonable supposition – many systemists have pointed out that systems are not only more than the sum of their parts but also *less* than the sum of their parts, due to parts being constrained by their systemic context. There is even a term for such loss or reduction of parts' powers: "submergence" (Bunge, 2003). A suggestion that emergence is accompanied by submergence is therefore not in itself new. However, the insight emerging here goes further in two important ways.

First, submergence is now *expected*, on principled grounds, rather than just being observed. This is called "retro-diction" in science, where we find a theoretical way of predicting something which was already known to be the case but only by observation, not via scientific arguments. Achieving retro-diction is an important step towards building a theory with predictive powers.

Second, because this claim is being made in a scientific way it can be empirically checked in a precise way. In effect we have replaced a qualitative aphorism, that emergence is accompanied by submergence, with a precise quantifiable scientific proposition, namely that the energy gained as the emergent stability property of the atom will be exactly matched by some kind of energy lost to the particles through property submergence.

This proposal is empirically testable, and when it is checked for atoms the expected result is indeed found: the mass of an atom is less than the sum of the

---

[9] Note that conservation principles can have time-dependencies, so that energy is conserved over an interval.

masses of its particles in their unbound states. This phenomenon is well known in nuclear chemistry, where it is called the "mass defect". Einstein's famous law $E = mc^2$ relates mass to energy, and we thus calculate an amount of energy this lost mass represents. When this is done it is indeed found to be the exact amount known as the "binding energy" of the atom, which is the energy that would be needed to break the atom up again.

Rousseau has proposed (Rousseau, 2018a) that this illustrates a general principle applicable to all systems, and hence to call it the 'Conservation of Properties Principle (CPP)'. CPP states that the energy associated with an emergent property in system formation is exactly matched by the sum of the energies lost by the parts participating in that systemizing interaction. More colloquially, this can be stated as "emergent properties are exactly paid for by submerged ones".

This principle presents a valuable insight for systems research, system design and systemic intervention. It provides an empirical standard for demonstrating that an observed system property is an emergent one, by connecting it with submergence. This is important because it casts suspicion on the common practice of calling any property noted at the system level but not seen in the parts an emergent one. CPP suggests that if the balancing interplay between emergence and submergence cannot be demonstrated, then the analysis is incomplete or wrong. For example, the boundary of the system may have been drawn incorrectly, and the supposedly emergent system-level behaviour is actually due to the action of parts unwittingly left out of consideration but which are in fact contributing that power to the whole in a summative way. Alternatively, the parts may have been mischaracterised, and have properties not currently attributed to them, and once again the system level property is actually summative rather than emergent. Either way research investigating the nature of a supposed emergent property will proceed differently from how this might be done without knowledge of CPP.

A further value is suggested via the idea that systems are dynamic structures, and so there is a constant interplay between emergence and submergence. This implies that when a system is suffering degradation due to loss of parts or weakening of inter-part interactions then we should be concerned not only about the loss of functionality but also about the re-emergence of previously inhibited behaviours of the remaining parts. This explains why it is so difficult to conserve or restore degraded or degrading complex systems (for example ecosystems). In systemic interventions both emergence and submergence have to be managed, and lack of control in this management might imply that the wrong boundaries have been managed, or the boundaries and/or parts have been mismanaged. In this way systemic interventions and also the design of resilient systems might now proceed differently from the way they would have been done without knowledge of CPP, and in particular this may help to reduce the occurrence or severity of unintended consequences.

It is not possible at this time to show that CCP applies across all systems types in the exact way the principle states, because we do not yet have a quantifiable scientific understanding of all the kinds of properties systems exhibit. This is especially notable in the case of living systems exhibiting mental or psychological properties. However, the principle does seem valid in a qualitative way, for example teams or families can achieve things the individual members cannot do by themselves, but members of such social units are also constrained in their behaviour compared to what they are able or willing to do in isolation. Some kind of balancing interplay seems to be in play here, as the willingness of an individual to accept constraints on their personal freedom seem to be dependent on the value they place on the benefits they gain though the powers of the social unit.

Although CPP cannot be applied in an exact way in such complex scenarios, because of the incompleteness of certain specialized sciences, CPP may still be useful in those contexts, because insights into systemic behaviour we gain by studying quantifiable cases could be translatable into metaphors providing effective new heuristic principles we can apply in more complex situations. As the sciences advance these metaphors can be improved, and become more scientific in the guidance they suggest.

6.4.2   Emergence and Super-Systems

By knowing the first systems principle we can immediately suggest another one, as follows. Systems hierarchy diagrams of the sort shown in Fig. 6.8 illustrate how system levels scale with size and complexity, but this somewhat obscures the fact that it represents a containment hierarchy, so that the systems at every level not only contain parts from the lower levels ("sub-systems") but are also themselves embedded as parts in higher-level systems ("super-systems"). A diagram showing the 'levels' as nested boxes, like a Russian doll, might illustrate this better than the stacked boxes in Fig. 6.8. Arthur Koestler coined the terms "holon" and "holarchy" to refer to systems and their inter-relationships regarded in this way (Koestler, 1967).

Holons and holarchies represent important systems ideas. A core concept of systems thinking is that things not only have environments but they are systemically connected into their environments, so every concrete thing short of the universe is a part in at least one super-system. In this light it is obvious that, in accordance with CCP, it must be that case that system properties are not only emergent over the properties of the parts, but are themselves subject to submergence as a result of their integration into their super-systemic context. This entails then that in fact systemic properties are determined by a balancing act between the bottom-up influence due to the parts and the outside-in influence of the super-systemic context. This provides a second systems principle, which Rousseau calls the "Principle of Universal Interdependence", and paraphrases as "system properties represent a balance between bottom-up emergence and outside-in submergence" (Rousseau, 2018a).

It is worth noting that this principle reflects a different idea from the statement often made for systems that they cannot be explained reductionistically because they involve an interplay between "bottom up" causation and "top-down" causation. That view is about how emergent properties can act back onto the parts, for example mental properties might emerge 'bottom up' via brain complexity can then influence processes within the body via will-power and bio-feedback. This kind of claim is not to do with a system's environment but is rather just a more sophisticated view about the goings-on within the system boundary.

The Principle of Universal Interdependence has significant implications for science. It entails that to model a system's real potential one has to look not only at what the parts contributed (bottom-up causation) but also what was deducted by the super-systemic context (out-side in causation). It means the explanatory arrows go both ways, both down and up from the system boundary. From a philosophy of science point of view this replaces classical "down-ward only" reductionism with a type of holistic interdependence perspective. For scientific research, this then suggests that for a theory about any new phenomenon the explanatory burden is expanded to now include both bottom-up and outside-in influences, and to do so in a balancing way. This principle also has significance for planning interventions and system designs, because it implies that there are two interconnected kinds of leverage points for changing system capacity/behaviour, namely via modulation of either the bottom-up or the outside-in influences.

In addition, this principle makes a contribution to epistemology, by adding a new theoretical virtue: theories and designs will be "better" if they are (more) holistic. An interesting prediction follows from this suggestion, namely that all the specialised disciplines will become more holistic as they mature. This is already happening in several fields, most notably at this time in cosmology, biology and medicine. It is therefore likely that a future systems engineering will not only be holistic itself (as the INCOSE Vision 2025 already calls for (International Council on Systems Engineering (INCOSE), 2014)) but will increasingly be able to draw on holistic specialised sciences for support.

6.4.3 Emergence and Complexity

The first two systems principles were derived by looking at interactions of low complexity. With these in hand we can now consider more complex cases.

Consider a super-system (W) consisting of two sub-systems, one of high complexity (S1) and one of low complexity (S2). The interactions between S1 and S2 bind them into the super-system (W). As a new system W has emergent new properties, and by the Conservation of Properties Principle (CPP) both S1 and S2 must undergo some degree of submergence. The binding interaction that links the two subsystems together is the same for each, but the relative impacts are unequal. A simple example will make this evident. Take for example the impact of gravitational attraction between a very small body and a large one, such as a meteoroid passing a planet. They form a system and each falls towards the other in accordance with Newton's Law of Gravity, but the effect on each is very different: the meteoroid's behaviour is strongly conditioned by the nearby planet, but the planet is hardly affected.

The interaction force is the same for each of the interaction partners, so it follows that they each give up the same amount of (gravitational potential) energy, so they contribute equally to the emergence of the new whole. In terms of CPP, we can say, speaking colloquially, that they each pay the same amount towards the emergent property of the whole, but the complex subsystem can afford that payment more easily, so is less affected by it. In a simple subsystem like S2 the few parts each have to give up a lot of their energy to make up their contribution to the total, but in a complex subsystem like S1 the many parts each give up a relatively small amount to make up their contribution. In line with the energy conservation aspect of CCP this conclusion can be generalised by saying that in systemizing interactions complex parts pay proportionately less towards emergent properties of the whole than simpler parts do. This amounts to a new systems principle, which Rousseau has called the "Principle of Complexity Dominance". It states that the impact of submergence on a part is proportional to the complexity differential between the part and the whole, and can be paraphrased as "complexity buffers autonomy".

This principle has relevance for scientific research, because it implies that when modelling the nature and potential of a given system the two explanatory arrows ('bottom up' and 'outside in') differ in weight in proportion to the relative complexity of the target system compared to the other systems making up the super-system it is systemically interlinked with. This is an important consideration in the study of naturalistic systems, because they cannot be completely shielded from systemizing interactions. This principle also applies to the behaviour and performance of designed systems, as they, like natural systems, are always parts in super-systems.

This principle is also relevant for planning systemic interventions, because the two inter-related leverage points for modulating system behaviour would be unequally weighted if there are complexity differentials involved. One corollary of this is that a target system can be efficiently controlled by a more complex one, as suggested by Ashby's so-called "Law of Requisite Variety".

Application of this principle however requires some care, because 'complexity' has multiple dimensions. The distinctions between these dimensions are far from sorted out, but we can separate them to some extent. Two important distinctions are between what we might call (for want of better terms) 'degree of complexity' and 'kind of complexity'. Systems are nature's way of creating complex enduring structures, and the two mentioned complexity dimensions reflect two aspects of nature's innovation process, one hallmarked by increases in scale and one hallmarked by increases in variety of behaviours. The two factors interdepend, with advances in the former (scale) often opening up opportunities for advances in the latter (behaviour).

We previously discussed how different kinds of systems can be grouped into a levels hierarchy, as shown in Fig. 6.2. This represents a type of complexity hierarchy, where the systems at each higher level have a new kind of behavioural property that emerges due to their higher level of organizational complexity. These levels represent not just an increase in complexity but shifts to new *kinds* of complexity. On this view biological systems thus appear 'higher up' in the system levels hierarchy than chemical systems because their increased behavioural variety is due to their having a radically different kind of complexity.

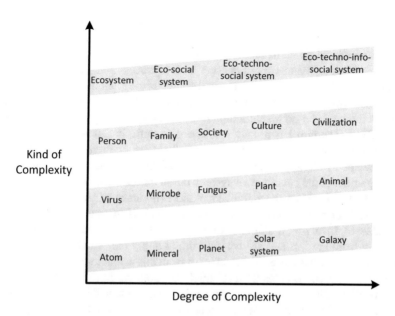

Fig. 6.10 Kinds and degrees of complexity in systems

However, there is also another aspect to complexity, which we previously encountered in Fig. 6.8. Here, on the left hand side we see a hierarchy of physical systems ranging from atoms to galaxies. This also represents some kind of complexity hierarchy, but here an increasing 'degree of complexity' enables the establishment of ever larger enduring structures by combining smaller assemblies of a similar kind in special ways. Large-scale systems of a certain kind are thus distinguished from small scale systems of the same kind by having a higher *degree* of complexity. This is the case for systems of all kinds, so we can illustrate the interplay between these two dimensions of complexity as shown in Fig. 6.10. Note that both dimensions of complexity are involved in the evolution of new system types, as is suggested by the sloping levels. An increase in scale does entail some increase in the level of behavioural complexity, but not of such a radically different kind as is required for producing a wholly new kind of system behaviour.

With this distinction in mind we can now see that in the example considered earlier, presented to expose the principle of complexity dominance, we only looked at a differential in the degree of complexity and not differentials in kinds of complexity. This might be taken to imply that the principle would only apply within system levels (that is, between systems sharing the same kind of complexity), but it can indeed be applied when interactions *across* system levels are at stake. However, in this case the principle has to be used carefully in order not to conflate the different dimensions of complexity in play. One way to do this is to consider kinds of complexity as conferring kinds of emergent properties on systems, and to recognize that kinds of interactions are exchanges between kinds of properties. We can thus treat complexity dominance as playing out relative to the emergent properties the interacting systems have in common, irrespective of the overall complexity of each of those systems.

For example, a bull is overall a more complex system than a fence, in the sense that the bull has a greater variety of parts, states, process and emergent properties. However, when the bull tries to get out of the field the relevant interaction is between the physical-system-level properties of each party, and here the fencing system outperforms the bovine system, entraining more mass and leverage into a more enduring structure. The bull has to stay inside or risk suffering more damage than the fence would in a forceful interaction between them. An intelligent bull might nevertheless escape by identifying and exploiting a weakness in the design, build quality or management of the containment system, but in this case the complexity differential involved is between the intelligence and experience of the bull, the fencing system designer and the farmer. This however involves interaction within a higher systems level, involving a different kind of complexity, so it has to be considered separately from the former case when trying to apply the principle of complexity dominance. The general lesson we can learn from this is that to fully model how the interaction between complex systems might turn out we have to identify, for that scenario, all the kinds of (emergent) properties between which interactions can take place, and identify for each possible interaction not only the interaction magnitude, activation probability and triggering conditions, but also the complexity differential. In such a model the principle of complexity dominance, applied across all these causal relationships, will give us insight into the overall 'possibility space' of the total interaction outcome, even if the kinds of systems involved are very different from each other.[10]

6.5   Summary and Future Prospects

We have in this chapter explored the nature of principles in general and systems principles in particular, and discussed how they relate to worldviews, laws theories and theoretical virtues. Using these ideas we presented three general scientific systems principles.

The three principles derived here are stated rather simply, and will benefit from further refinement and also formalization. Such work is presently under way, as are empirical studies to evaluate their use in practice (Calvo-Amodio et al., 2017).

---

[10]The example just given, of a fence containing a bull, could mistakenly be taken as a counter-example to Ashby's "Law of Requisite Variety", which is often loosely stated as that to be effective a controller must be more complex (in the sense of having more variety) than the system it controls. However, careful framings of Ashby's Law are stated in terms of degrees of freedom, and then specifically claim that *relative to the process being controlled* the degree of freedom of the controller must be larger than the degree of freedom of the process being controlled. This is a paraphrase of the argument developed here for only applying a rule relative to a specific dimension of interaction. Under Ashby's Law, an effective control system must therefore have more variety relative to each of the kinds of processes to be regulated in the controlled system.

More importantly, their derivation has demonstrated the value of the principled foundations laid down in the GSTD program, which clarified relevant terminology of Systemology and showed the parallels between methods and models in science and philosophy of science more broadly and the methods and models the general systems movement is trying to derive. The three principles discussed here may be just the vanguard of a range of general scientific systems principles to come. In a recent work, Rousseau, drawing on the foundations laid by the GSTD programme, outlined nine strategies for discovering GSSPs (Rousseau, 2017a), one of which was followed to produce the three principles discussed in this chapter. This gives grounds for optimism that more GSSPs will soon follow. If this potential can be realised then the GST* the founders of the ISSS called for may be a realistic prospect for the near future, opening the door to the systematic discovery of further Systemics, and a general expansion in the significance of systems science for our ability to understand our world and translate our progress across communities. In this way General Systemology and the Translational Systems Science it supports could help transform our prospects for creating the 'better world' the pioneers of systems science envisioned.

References

Ackoff, R. L. (1971). Towards a system of systems concepts. *Management Science, 17*(11), 661–671.

Arnold, R. D., & Wade, J. P. (2015). A definition of systems thinking: A systems approach. *Procedia Computer Science, 44*, 669–678.

Augustine, N. R. (1987). *Augustine's Laws*. New York: Penguin.

Award#1645065 – EAGER/Collaborative Research: Lectures for Foundations in Systems Engineering. (n.d.). Retrieved January 15, 2017, from https://www.nsf.gov/awardsearch/showAward?AWD_ID=1645065&HistoricalAwards=false

Boulding, K. E. (1956). General systems theory – The skeleton of science. *Management Science, 2*(3), 197–208.

Bunge, M. (1979). *Ontology II: A world of systems*. Dordrecht: Reidel.

Bunge, M. (2000). Energy: Between physics and metaphysics. *Science & Education, 9*(5), 459–463.

Bunge, M. (2003). *Emergence and convergence: Qualitative novelty and the unity of knowledge*. Toronto: University of Toronto Press.

Calvo-Amodio, J., Kittelman, S., & Wang, S. (2017). *Applications of systems science principles*. Presented in the Workshop on 'Systems Science – Principles for the Evolution of Systems', held on the 18th of October 2017 as part of the 2017. International annual conference of the American Society for Engineering Management (ASEM) on the theme 'Reimagining Systems Engineering and Management', October 18th–21st, 2017, Marriott Hotel – Huntsville, Alabama, USA.

Chakravartty, A. (2010). *A metaphysics for scientific realism: Knowing the unobservable*. Cambridge, UK: Cambridge University Press.

Chakravartty, A. (2013, Summer). Scientific realism. In E. N. Zalta (Ed.), *The stanford encyclopedia of philosophy*. Retrieved from http://plato.stanford.edu/archives/sum2013/entries/scientific-realism/

Collopy, P. D., & Mesmer, B. L. (2014). *Report on the Science of systems engineering workshop*. In 53rd AIAA Aerospace Sciences meeting (pp. 1–4). Kissimmee, FL: American Institute of Aeronautics and Astronautics.

Danielle-Allegro, B., & Smith, G. (2016). *Exploring the branches of the systems tree*. Castelnau d'Estretefonds: Les editions Allegro Brigitte D.

De Caro, M., & Macarthur, D. (2010). *Naturalism and normativity*. New York: Columbia University Press.

Dilworth, C. (1996). *The metaphysics of science: An account of modern science in terms of principles, Laws and Theories*. Dordrecht: Springer.

Ellis, B. (2009). *The metaphysics of scientific realism*. Durham: Routledge.

Flyvbjerg, B. (2014). What you should know about megaprojects and why: An overview. *Project Management Journal, 45*(2), 6–19.

Francois, C. (Ed.). (2004). *International encyclopedia of systems and cybernetics*. Munich: Saur Verlag.

Friendshuh, L., & Troncale, L. R. (2012). *Identifying fundamental systems processes for a general theory of systems*. In Proceedings of the 56th Annual conference, International Society for the Systems Sciences (ISSS), July 15–20, San Jose State University, 23 pp.

Gall, J. (1978). *Systemantics: How systems work and especially how they fail*. New York: Pocket Books.

Gillett, C., & Loewer, B. (Eds.). (2001). *Physicalism and its discontents*. New York: Cambridge University Press.

Hardy-Vallee, B. (2012). The cost of bad project management. *Business Journal (Gallup Inc)*. Retrieved from http://www.gallup.com/businessjournal/152429/Cost-Bad-Project-Management.aspx

Heylighen, F. (2012). Self-organization of complex, intelligent systems: An action ontology for transdisciplinary integration. *Integral Review*. Retrieved from http://pespmc1.vub.ac.be/papers/ECCO-paradigm.pdf

Hitchins, D. K. (1992). *Putting systems to work*. Chichester, UK: Wiley.

International Council on Systems Engineering (INCOSE). (2014). *A world in motion – Systems engineering vision 2025*. San Diego, CA: INCOSE. Retrieved from http://www.incose.org/AboutSE/sevision

Jones, D. (2016, February 20). *66% of IT Projects fail*. Retrieved April 27, 2017, from https://projectjournal.co.uk/2016/02/20/66-of-it-projects-fail/

Katina, P. F. (2016). Systems theory as a foundation for discovery of pathologies for complex system problem formulation. In*Applications of systems thinking and soft operations research in managing complexity* (pp. 227–267). Cham: Springer. https://doi.org/10.1007/978-3-319-21106-0_11.

Koestler, A. (1967). *The ghost in the machine*. Chicago: Henry Regnery.

Koons, R. C., & Bealer, G. (Eds.). (2010). *The waning of materialism*. Oxford: Oxford University Press.

Kramer, N. J. T. A., & De Smit, J. (1977). *Systems thinking: Concepts and notions*. Leiden: Martinus Nijhoff.

Krigsman, M. (2012, April 10). *Worldwide cost of IT failure (revisited): $3 trillion*. Retrieved November 27, 2017, from http://www.zdnet.com/article/worldwide-cost-of-it-failure-revisited-3-trillion/

Lineberger, R. (2016, October 24). *Deloitte study: Cost overruns persist in major defense programs — Press release | Deloitte US*. Retrieved February 27, 2017, from https://www2.deloitte.com/us/en/pages/about-deloitte/articles/press-releases/cost-overruns-persist-in-major-defense-programs.html

Maier, M. W., & Rechtin, E. (2009). *The art of systems architecting* (3rd ed.). Boca Raton, FL: CRC Press.

Matthewson, J., & Weisberg, M. (2009). The structure of tradeoffs in model building. *Synthese, 170*(1), 169–190.

Maxwell, N. (2004). Non-empirical requirements scientific theories must satisfy: Simplicity, unification, explanation, beauty. In J. Earman & J. Norton (Eds.), *PhilSci archive*. Pentire Press. Retrieved from http://philsci-archive.pitt.edu/1759/

McNamara, C., & Troncale, L. R. (2012). *SPT II.: How to find & map linkage propositions for a general theory of systems from the Natural Sciences literature*. In Proceedings of the 56th Annual Conference, International Society for the Systems Sciences (ISSS), July 15–20, San Jose State University, 17 pp.

Mingers, J. (2006). *Realising systems thinking: Knowledge and action in management science*. New York: Springer.

Mingers, J. (2014). *Systems thinking, critical realism and philosophy: A confluence of ideas*. New York: Routledge.

Mobus, G. E., & Kalton, M. C. (2014). *Principles of systems science* (2015th ed.). New York: Springer.

Poli, R. (2001). The basic problem of the theory of levels of reality. *Axiomathes, 12*(3–4), 261–283. https://doi.org/10.1023/A:1015845217681.

Pratt, J. M., & Cook, S. C. (2018). A principles framework to inform Defence SoSE Methodologies. In A. M. Madni, B. Boehm, R. Ghanem, D. Erwin, & M. J. Wheaton (Eds.), *Disciplinary convergence in systems engineering research* (pp. 197–213). Cham: Springer.

Rapoport, A. (1968). General system theory. In D. L. Sills (Ed.), *The International encyclopedia of Social Sciences* (Vol. 15, pp. 452–458). New York: Macmillan/The Free Press.

Rousseau, D. (2011). *Minds, souls and nature: A systems-philosophical analysis of the mind-body relationship in the light of near-death experiences* (PhD thesis). University of Wales Trinity Saint David, Lampeter, Wales, UK.

Rousseau, D. (2017a). Strategies for discovering scientific systems principles. *Systems Research and Behavioral Science, 34*(5), 527–536. https://doi.org/10.1002/sres.2488.

Rousseau, D. (2017b, March 23–25). *Three general systems principles and their derivation: Insights from the philosophy of science applied to systems concepts*. In Proceedings of the 15th annual conference on systems engineering research – Disciplinary convergence: Implications for systems engineering research.

Rousseau, D. (2018a). Three general systems principles and their derivation: Insights from the philosophy of science applied to systems concepts. In A. M. Madni, B. Boehm, R. Ghanem, D. Erwin, & M. J. Wheaton (Eds.), *Disciplinary convergence in systems engineering research* (pp. 665–681). New York, NY: Springer. https://doi.org/10.1007/978-3-319-62217-0_46.

Rousseau, D. (2018b). On the architecture of systemology and the typology of its principles. *Systems, 6*(1), 7. https://doi.org/10.3390/systems6010007.

Rousseau, D., Wilby, J. M., Billingham, J., & Blachfellner, S. (2015, August 4, 2015). *Manifesto for General Systems Transdisciplinarity (GSTD)*. Plenary Presentation at the 59th Conference of the International Society for the Systems Sciences (ISSS), Berlin, Germany, see also http://systemology.org/manifesto.html.

Rueger, A., & McGivern, P. (2010). Hierarchies and levels of reality. *Synthese, 176*(3), 379–397. https://doi.org/10.1007/s11229-009-9572-2.

Schindel, W. D. (2011). *What is the smallest model of a system?* In Proceedings of the INCOSE 2011 International symposium, International Council on Systems Engineering (Vol. 21(1), pp. 99–113). Citeseer.

Schindel, W. D. (2013). *Abbreviated Systematica™ 4.0 glossary—Ordered by concept*. Terre Haute, IN: ICTT System Sciences.

Schoderbek, P. P., Schoderbek, C. G., & Kefalas, A. G. (1990). *Management systems: Conceptual considerations* (Revised ed.). Boston: IRWIN.

Schöttl, F., & Lindemann, U. (2015). Quantifying the complexity of socio-technical systems–a generic, interdisciplinary approach. *Procedia Computer Science, 44*, 1–10.

Schrenk, M. (2016). *Metaphysics of science: A systematic and historical introduction*. New York: Routledge.

Senge, P. M. (1990). *The fifth discipline: The art and practice of the learning organization*. London: Random House.

Sillitto, H., Dori, D., Griego, R. M., Jackson, S., Krob, D., Godfrey, P., et al. (2017). Defining "system": A comprehensive approach. *INCOSE International Symposium, 27*(1), 170–186. https://doi.org/10.1002/j.2334-5837.2017.00352.x.

Sillitto, H., Dori, D., Griego, R. M., Krob, D., Arnold, E., McKinney, D., Godfrey, P., Martin, J., Jackson, S. (2018). What do we mean by "system"? - System beliefs and worldviews in the INCOSE community. Proceedings of the INCOSE International Symposium, Washington, DC, USA, 7–12 July 2018. (17 pages).

Syal, R. (2013, September 18). Abandoned NHS IT system has cost £10bn so far. *The Guardian*. Retrieved from https://www.theguardian.com/society/2013/sep/18/nhs-records-system-10bn

Troncale, L. R. (1978). Linkage propositions between fifty principal systems concepts. In G. J. Klir (Ed.), *Applied general systems research* (pp. 29–52). New York: Plenum Press.

Troncale, L. R. (1985). The future of general systems research: Obstacles, potentials, case studies. *Systems Research, 2*(1), 43–84.

Troncale, L. R. (1988). The systems sciences: What are they? Are they one, or many? *European Journal of Operational Research, 37*(1), 8–33.

von Bertalanffy, L. (1950). An outline of general system theory. *British Journal for the Philosophy of Science, 1*(2), 134–165.

von Bertalanffy, L. (1956). General System Theory. *General Systems [Article Reprinted in Midgley, G. (Ed) (2003) 'Systems Thinking' (London: Sage) Vol 1, pp 36–51. Page Number References in the Text Refer to the Reprint.], 1*, 1–10.

von Bertalanffy, L. (1969). *General system theory: Foundations, development, applications*. New York: Braziller.

Walden, D. D., Roedler, G. J., Forsberg, K. J., Hamelin, R. D., & Shortell, T. M. (Eds.). (2015). *Systems engineering handbook* (4th ed.). Hoboken, NJ: Wiley.

Weinberg, S. (2008). Newtonianism, reductionism and the art of congressional testimony. In M. A. Bedau & P. Humphreys (Eds.), *Emergence: Contemporary readings in philosophy and science* (pp. 345–357). Cambridge, MA: MIT Press. Retrieved from http://ip144.qb.fcen.uba.ar/libroslfp/Historia%20y%20Filosofia%20de%20la%20Ciencia/Bedau%20-%20Emergence%20-%20Contemporary%20Readings%20in%20Philosophy%20and%20Science%20(MIT,%202008).pdf#page=362

Wilby, J. M., Rousseau, D., Midgley, G., Drack, M., Billingham, J., & Zimmermann, R. (2015). Philosophical foundations for the modern systems movement. In M. Edson, G. Metcalf, G. Chroust, N. Nguyen, & S. Blachfellner (Eds.), *Systems thinking: New directions in theory, practice and application*, Proceedings of the 17th Conversation of the International Federation for Systems Research, St. Magdalena, Linz, Austria, 27 April–2 May 2014 (pp. 32–42). Linz: SEA-Publications, Johannes Kepler University.

Wimsatt, W. (2007). *Re-engineering philosophy for limited beings: Piecewise approximations to reality*. Cambridge, MA: Harvard University Press.

Index

Printed in the United States
By Bookmasters